IN PURSUIT OF THE MOON

THE HUNT FOR A MAJOR NASA CONTRACT

BILL TOWNSEND

IN PURSUIT OF THE MOON
THE HUNT FOR A MAJOR NASA CONTRACT

This is a work of nonfiction based on the author's memory of the events as they had unfolded. Any errors in timing or factual descriptions of events are purely unintentional. The story as told represents the author's best efforts to tell a true story of the actual events that occurred during this pursuit, and the people associated with those events.

Because of the constantly changing nature of the websites that were used to obtain selected material, the author has updated the associated citations to be as current, accurate, and complete as possible. However, in some cases, the material in question was no longer available on the website from which it had originally been obtained. For those situations, the author has identified alternate websites for the material, to the maximum extent practical.

iUniverse books may be ordered through booksellers or by contacting:

iUniverse
1663 Liberty Drive
Bloomington, IN 47403
www.iuniverse.com
1-800-Authors (1-800-288-4677)

Any people depicted in stock imagery provided by Getty Images are models, and such images are being used for illustrative purposes only. Certain stock imagery © Getty Images.

Cover Picture Credit: NASA (Public Domain) via Wikimedia Commons, Ares-I Launch, February 2, 2008, https://upload.wikimedia.org/ wikipedia/commons/8/8d/Ares-1_launch_2_02-2008.jpg.

ISBN: 978-1-5320-7916-0 (sc)
ISBN: 978-1-5320-7929-0 (e)

Library of Congress Control Number: 2019911921

Print information available on the last page.

iUniverse rev. date: 11/14/2019

This book is dedicated to my wonderful wife of more than fifty years, Carolyn, who has stood by, supported, and advised me in all matters related to my career with NASA and, subsequently, with Ball Aerospace—not to mention raising our children in an excellent fashion while I pursued that career. Furthermore, given her own thirty-five-year career with NASA, she understood and appreciated very well the many challenges, constraints, and nuances of the line of business we were in, which was immensely helpful to us both. And relative to the subject of this book, she was absolutely crucial to Ball Aerospace's Ares I Instrument Unit Avionics (IUA) pursuit activities in Huntsville, Alabama. As our community liaison, she quickly got us integrated into the Huntsville business community, so much so that Ball Aerospace was referred to by the *Huntsville Times* as a "local firm" only six months after our arrival!

Contents

Acknowledgments

Many, many people deserve recognition for their contributions to the significant procurement activity described herein. While only a few of the many involved will be specifically mentioned, there were obviously very many other key contributors:

- Certainly at the top of the list is my boss, Dave Taylor, then president and CEO of Ball Aerospace and a steadfast supporter of Ball Aerospace's pursuit of the Ares I IUA procurement. Without his leadership and support, this effort would never have been possible.
- Dave Murrow, the capture manager, kept the many odd-size nuts and bolts of this very complicated pursuit screwed together properly, and he was instrumental in securing the support of our excellent teammates.
- Bill Unger, the chief financial officer, and Jim Stevens, the VP for human resources, were key to making their departments available not only for the pursuit but also for the development activity if we were to win.
- Sarah Sloan, as head of the communications group, supported, fiscally and otherwise, the many outreach activities that took place in Huntsville, Alabama.
- Vickie Terry and her team did an incredible job of putting our various proposals together, always at the last possible minute (very sorry, Vickie).

- Parker Counts, our on-site consultant, knew the NASA Marshall Space Flight Center (Marshall) inside and out and thus advised us well as to the expectations of our counterparts on the Marshall side.
- Marlene Kruise, my administrative assistant in Boulder, saw to it that I was always able to go wherever and whenever I was needed to go. She made many other significant contributions too numerous to mention.
- Carolyn Townsend, our community liaison, did a truly outstanding job of quickly integrating Ball Aerospace into the local Huntsville and NASA Marshall communities.
- And too many others to name individually who were either working on-site in Huntsville, Alabama, or working back at our home base in Boulder, Colorado, but thank you to all who made contributions, large or small, to this effort.

Thanks so much to each and every one of you for your active involvement and many contributions. Without your collective efforts, this pursuit would never have resulted in a selectable proposal, something that seemed a real stretch when we started.

Introduction

Having spent more than fifty years working in the aerospace industry, the first forty with NASA, I had always thought I would write a book about everything I had seen and done during the course of my career. However, at the end of the day, I elected to write a book about a singular activity that occurred—one that was incredibly challenging, that I enjoyed immensely, and that left an indelible mark on me, a mark that remains to this day, a decade later.

This is the story of how I took thirty-five top-notch Birkenstock-wearing engineers (eventually growing to an on-site team of about sixty-five) from Ball Aerospace in Boulder, Colorado, to the cosmopolitan southern city of Huntsville, Alabama (better known in some circles as Rocket City USA). Once there, my wife, Carolyn, and I worked together toward a common goal: pursuing and winning a major new NASA contract by seeking to overcome the entrenched incumbent aerospace contractors who had dominated NASA's human space flight program for decades. Since Ball Aerospace was less than a tenth the size of its competitors, this is very much a David-versus-Goliath story. It's also an intriguing story about NASA and how it selected contractors to implement its exploration vision.

The purpose of this major competition was for NASA to select a contractor to build the brains of the new launch vehicle that would replace the space shuttle and take Americans back to the moon for the first time in forty years. NASA had communicated verbally and in writing certain things that they were looking for in the contractor they would select.

These attributes included new blood, innovation, and creativity. And because of that, three companies, including Ball Aerospace, each new to NASA's human space flight program, at least on this scale, decided to throw their hat into the ring. Add to that two deeply entrenched multibillion-dollar-a-year incumbent aerospace contractors for a total of five companies chasing this job. And because of what NASA was saying, the three new companies elected to spend many millions of dollars trying to grab the gold ring.

To facilitate our work in Huntsville, Carolyn and I were given a mostly clean sheet of paper to work with. As a consequence, we were able to do some really significant things to help position Ball Aerospace for a win. These included getting rapidly integrated into the local business community (thanks to Carolyn), selecting near-perfect teammates, and submitting a high-quality proposal that was way more than good enough to be selected by NASA.

This is not a theoretical book but one that takes you inside the inner workings of a real-life industry pursuit so that you can see, up close and personal, the excitement and the frustrations of such a pursuit. You will also see how a major government agency, NASA, behaves and conducts itself with its contractor base—sometimes not quite in the manner you might have expected.

PART 1

The Situation

The Trip to Washington, DC

My wife, Carolyn, and I had an early flight to Dulles Airport near Washington, DC, so we were up well before five in the morning. We had eight bags with us, as we were taking presents with us, in order to be able to celebrate Christmas with our kids on the East Coast. Since I didn't know exactly what was going to happen, or when we would next be in Colorado, Carolyn and I had taken our administrative assistants out the day before for a Christmas lunch to thank them for all their support during this long campaign.

It was a typical December (2007) morning: cold but with a clear sky. Colorado, deservedly so, is known for its snow. It is not as well known for its sunshine, yet Boulder, where Ball Aerospace is located, gets more than three hundred days of sunshine a year, directly rivaling places like San Diego, which are well known for their sunshine. Since we were headed to DC, we wouldn't get to see how things turned out this day, but from a long history of experience, I suspected we were about to miss yet another beautiful day in Boulder.

Making this trip were Dave Taylor (president and CEO of Ball Aerospace—and my boss); Dave Murrow (the Ares I Instrument Avionics

Unit (IUA) capture manager); Roz Brown (our media person); Carolyn (our community liaison); and myself (Ball Aerospace's VP for exploration systems).

Dave Taylor was in a good mood. I was not. Dave asked me several times if I was feeling okay, and I just said I was apprehensive about the outcome of all this. But I think I was the only one who actually thought there could be any other outcome than a win for Ball Aerospace. Certainly, if everything that we had been told multiple times by multiple people at NASA was true, we should be the winner. We had been told clearly that NASA wanted new blood, didn't want the same contractor for both the Ares I Upper Stage and the IUA, wanted creativity, wanted innovative thinking, and wanted to learn from the aircraft avionics world. So with the right proposal, we should be a shoo-in, right? But I wasn't sure, especially given all the concerns that I'd had all along about Boeing. In any event, we were about to learn the answer.

We landed at Dulles Airport, picked up our luggage, got into our cars, and headed to Carol Lane's (Ball Aerospace's director of Washington operations) DC office located in Arlington, Virginia. We arrived that afternoon around one. I had been told to expect the phone call from Doug Cooke (the NASA source selection official) between two and three o'clock. Since the call was coming to my cell phone, I could have taken it from anywhere, but if we won, NASA wanted our smiling faces to show up at the planned four o'clock press conference at NASA Headquarters, which was why we had flown to DC. Arrangements had already been made to speed us down to NASA Headquarters if we won.

We were all gathered in Dave's DC office, and most of us were sitting around the round table in front of his desk. The mood in the room was light and upbeat, and I had perked up quite a bit.

A little before two, I set my BlackBerry on the table, making sure it was not on vibrate and that the ringer volume was turned up. We even placed a call to it to make sure it was working properly.

Then we watched and waited.

The Vision for Space Exploration

So how did we get to this place at this time? To better understand that part of the story, let's fast rewind to three in the afternoon on January 14, 2004, at NASA Headquarters in Washington, DC, as an unprecedented event was about to take place. The forty-third president of the United States, George W. Bush, was coming to headquarters to announce what would soon become known as the President's Vision for Space Exploration. Later, as people began to realize the political liability of having the Vision directly attached to President Bush, whose popularity was falling precipitously because of the war in Iraq, the realization that there were no weapons of mass destruction (WMDs), the never-ending hunt for Osama bin Laden, and so on, it was shortened to the Vision for Space Exploration. Still later it became known simply as the Vision, and within NASA it became known even more simply as Exploration.

But on this day, it was President Bush's Vision for Space Exploration, and I was there to see it unveiled. NASA had long needed a new direction, something NASA had been searching for, ever since the Apollo program ended. The current NASA administrator, Sean O'Keefe, had been working diligently behind the scenes with Vice President Dick Cheney and many others in the White House, the Office of Management and Budget (OMB), and the Office of Science and Technology Policy (OSTP) to craft this new vision. He could do that because unlike most recent NASA administrators, he had come to NASA from OMB and had good connections directly into the White House. Now was the time to publicly unveil the new plan to the rest of NASA and to the world.

The vision that the president presented that day was disarmingly simple. Humans would return to the moon and then go on to Mars. But it came with no additional money, and there was no real timetable. Rather, NASA was being told to execute the new plan within its existing budget, plus inflation. I thought right away that this plan was doomed to failure, but who was I to be telling that to the president?

What happened next was a form of carnage. The word was passed down that if it isn't "Exploration," it isn't NASA. Thus, the rush was on within NASA to steer everything possible toward being essential to Exploration. Of particular note was that the science programs at NASA were threatened with annihilation if they failed to get on board by making sure their research activities also supported Exploration. And NASA's aeronautics program, the A in NASA, which had been the very basis upon which NASA was established in 1958, was clearly on the cutting-room floor.

As the deputy center director (the number-two person) of NASA's Goddard Space Flight Center, an institution at whose heart is a vibrant science program, I felt severely disenfranchised by the new "Vision." I wasn't alone. My boss, Al Diaz, the center director, and I, found ourselves in deep discussions with our senior staff about how best to respond to this sudden change of direction for the agency. We were already working feverishly to refocus everything possible on Exploration, but what else could we do? One such set of discussions led to us reaching out to NASA's Jet Propulsion Laboratory (JPL) in Pasadena, California, the only other NASA center that was focused almost totally on robotic space flight. Over the years, Al and I had worked hard to collaborate with JPL wherever it made sense to do so (for example, where our respective capabilities and ideas could complement and strengthen one another), and JPL had reciprocated in a similar fashion. This process of trying to work better together had been formalized into regular executive retreats so that we could better get our entire management teams on board with whatever was the latest thinking as to how we could best work together.

We began talking again but with much more urgency than usual. Out of these discussions came an agreement as to how best to work together in support of the new Exploration program. The agreement that we finally reached was that Goddard would build the robotic orbiting spacecraft required to support the Exploration program at the moon and JPL would build the robotic landers (i.e., those spacecraft that would actually touch down on the moon's surface as precursors to humans doing so [again]). The arrangement for working together on Mars orbiting spacecraft and landers was to be determined, as that was not the issue of the day. This division of responsibilities made a lot of sense, as it played to each institution's strengths. It also was an agreement that was quietly endorsed by the NASA Headquarters Office of Space Science.

So, and not without having thought through this very carefully, as the Vision began to take better shape, Goddard found itself with an immediate role: to build the very first moon-orbiting spacecraft required to begin the search for the best landing sites on the moon (where humans would ultimately set foot once again, after not having done so for more than thirty years). This mission became known as the Lunar Reconnaissance Orbiter (LRO), which was successfully launched and inserted into lunar orbit in June 2009. Similarly, JPL was getting itself well positioned on the landing side of the Exploration program. And the respective science programs, under NASA Headquarters' leadership, were also being realigned to focus on Exploration style science that was directed at better executing NASA's new Vision.

In the meantime, the reality of the Vision was becoming more and more apparent. The problem was that there wasn't enough money to make anything significant happen on any reasonable timescale. Nor was there very much money to pursue the development of new technology to help facilitate getting back to the moon, and on to Mars, both more efficiently and more safely. And there were many serious hurtles to clear in this regard as well, not the least of which was dealing with the effects of the radiation environment on humans and figuring out how to get back from Mars once there. But NASA has always loved a challenge, so that was all just fine, except for the lack of funding to get these things executed properly. Thus, compromise became the keyword of the day.

Other complicating factors included, simultaneous with all of the above, having (needing) to respond to all the recommendations and findings of the Columbia Accident Investigation Board (CAIB) report (released in stages from August to October 2003), in the aftermath of the February 1, 2003, space shuttle *Columbia* accident.

But in the meantime, other things were happening that would change the course of events for me.

My Move to Ball Aerospace

Thursday, June 24, 2004, almost six months after President Bush's announcement of his Vision for Space Exploration, had been quite a day for me. My boss, Al Diaz, the center director of Goddard, located in Greenbelt, Maryland, just outside the Washington, DC, beltway, had come home early from his planned monthlong vacation in Italy with his wife, Angela, and some friends. As the deputy center director for more than six years, I had been the acting center director in Al's absence. Frankly, I was initially glad to see him come back early, as I was hoping that I could take some time off too to go sailing on the Chesapeake Bay with my family, but that was not to be. Once I caught Al up on things that had been going on while he was gone, he told me why he had come home early. Sean O'Keefe, the NASA administrator at the time, had asked Al to return to NASA Headquarters in DC to take over the helm of the new Science Mission Directorate (SMD) that had just been established by merging the Office of Earth Science (which I had previously led for almost two years) with the Office of Space Science. Not only that, but Al intended to take Alison McNally, associate director of Goddard, and thus, the number-three person at Goddard, with him to NASA Headquarters. Furthermore,

Ed Weiler, previously the associate administrator of the Office of Space Science, which had been merged into the new SMD, would be taking over the helm of Goddard as its tenth center director. All of this would be effective August 1, except that Al would be going downtown immediately with Alison, yet Ed Weiler would not be coming out to Goddard until August 1. So much for our sailing vacation.

That evening, as I returned to our home in Annapolis, Maryland, late as usual, the phone rang before I could even put my briefcase down. It was to be the call that changed my life. On the other end was an industry headhunter who wanted to talk to me about a job opportunity at Ball Aerospace & Technologies Corp. in Boulder, Colorado. NASA had done business with Ball Aerospace for many, many years, so I knew them well, going all the way back to my time at NASA Headquarters. The job in question was to be the vice president and general manager of their Civil Space Systems (CSS) Business Unit.

After a brief discussion with the headhunter, I told her I would need some time to think about this before telling her whether I was willing to throw my hat into this ring. So, as so often happens with these sorts of opportunities, this became the major topic of discussion with my wife of almost thirty-seven years, Carolyn, at dinner. The next day was my fifty-eighth birthday, and that evening, the headhunter called again. I told her I was interested in the position but that I would need to recuse myself at Goddard from all matters involving Ball Aerospace. She said she totally understood and had expected me to say just that.

The next morning, a Saturday, I called Larry Watson, the Goddard general counsel, and told him all this, and then asked him to draw up the necessary paperwork for me to hand off all Ball Aerospace matters to Dolly Perkins, the director of the Flight Programs and Projects Directorate at Goddard. Over the weekend, we all agreed to meet at seven thirty Monday morning, to formally effect this transfer of responsibility, which we did. In the meantime, Dave Taylor, the CEO and president of Ball Aerospace, had been planning to meet with me on Monday afternoon concerning some of the work that Ball Aerospace was doing with Goddard. Since I was now not able to have such a meeting with Dave, the purpose of his trip to Goddard changed to doing an interview with me.

Late Monday afternoon, I took a couple of hours off, drove over to nearby BWI Airport, and met with Dave in a conference room at the Signature General Aviation terminal for about an hour. As I walked into the terminal, who did I see but a group of Goddard folks who were just coming in from a trip to the NASA site at Wallops Island, Virginia (where I had worked for eighteen years earlier in my NASA career). I'm sure they wondered what was up, as it didn't appear that I was headed anywhere myself. In any event, Dave and I had a great discussion, and the ball, no pun intended, got rolling very quickly.

All was quiet until Thursday evening, July 1, when I got a call from the Ball Aerospace head of human resources, Joe Winslow. He wanted to know if I could come to Colorado the next morning to interview with Dave Hoover, the CEO and president of Ball Corporation, the parent company of Ball Aerospace, and the company that used to make Ball canning jars, but was now a major global packaging company doing some $7 billion a year in sales. As it turned out, I had to go to the West Coast on Monday, July 5, the "official" (for government employees) Fourth of July holiday, for the upcoming launch of the Earth Observing System (EOS) Aura spacecraft from Vandenberg Air Force Base, where I would be the senior Goddard person responsible for the launch and early on-orbit operations. Typically, this sort of trip would take me away from Goddard for a week to ten days, so I had arranged to take Friday, July 2, off so I could have some additional time with my family before being gone for a while. Thus, fortuitously, I was available on Friday without upsetting the apple cart at Goddard. So I said yes to the trip.

The next morning, I caught the six o'clock flight to Denver, where I was picked up by Joe and taken to Ball Corporation Headquarters in Broomfield, Colorado, to do my interviews (I also interviewed with David Westerlund, executive vice president, administration). After which, Joe took me back to the Denver airport, and by seven that same evening I was standing in our kitchen in Annapolis talking to Carolyn about the experience that I had just had. It had been another fascinating day.

On Monday, July 5, I left for California as planned for the EOS Aura launch (I later wrote an article about this launch titled "No Launch before Its Time,"[1] which is a great indicator of the stress that exists for any launch

---

[1] Bill Townsend, "No Launch Before Its Time," accessed April 29, 2019, https://appel.nasa.gov/2004/11/01/no-launch-before-its-time/.

of a major national asset), and three days later, exactly two weeks after the original headhunter call, I got an offer from Ball Aerospace for the job. Shortly after I gave Ball Aerospace a verbal yes, with the caveat that NASA legal still had to work out what my legal restrictions were going to be, given my past involvement with the aerospace industry, and especially with Ball Aerospace, as a government employee. Oh, by the way—the EOS Aura spacecraft was successfully launched on July 15, 2004.

As luck would have it, the activity to determine legally what I could and couldn't do after leaving NASA was not too far behind the Boeing/ Darleen Druyun government ethics fiasco (she ended up being sentenced to nine months in jail), so figuring out what to do with me took a lot longer than I would have liked. But finally, Goddard legal produced a twenty-one-page document of (mostly) don'ts, and I presented that to Ball Aerospace legal to see if they could live with all the restrictions being placed on my future interactions with the government, and especially NASA. Fortunately, Dave Taylor is a most honorable person, and he stuck by me through this despite the fact that NASA was tying both my hands behind my back, at least for a while.

On Sunday evening, August 8, I got a call from Joe Winslow at Ball Aerospace that all was clear for me to assume my new position with them effective Monday, September 13 (per my request, I wanted to give NASA a month's notice). The next morning, I went in early to meet with Ed Weiler, my new boss of only a week, to tell him that I would be leaving NASA. Having just assumed his new job, and never having run a NASA field center before, let alone one with almost ten thousand on-site/near-site employees, and an almost $3 billion budget, he was less than enthusiastic about my impending departure.

But the die was cast, and I was now on my way to Ball Aerospace, and a whole new set of challenges, as well as a continuation of the exciting work that I been doing in the space arena for more than forty years. I was given a very nice going-away party by Goddard, which Carolyn and I flew our son, Jason, in for, from Colorado (he was a student at the University of Colorado in Boulder).

Then, on Friday, September 10, Carolyn and I were winging our way to Colorado so that I could start a new career. Carolyn spent the next week helping me get set up in the apartment that Ball Aerospace had temporarily made available to me, and on Thursday evening of that week, Carolyn and I hosted the Ball Aerospace executive team at the apartment for a little "get to know everyone" after-work party. We also had flown our daughter, Tiffany, in so that she could see where Dad would be living and working, so both she and Jason were there, as well as Carolyn. On Sunday, September 19, I put Carolyn and Tiffany on a plane back to the East Coast, as Carolyn was staying back east for Tiffany's senior year of high school. Thus, for the first time in my adult life, I was alone at the beginning of a really big adventure. Fortunately, I was able to get back east every two to three weeks that first year, so that helped immensely.

One of the very first things I did in my new job was establish a regular series of all-hands meetings with everyone at Ball Aerospace who supported the programs that I was responsible for. The first meeting was during my first week at Ball Aerospace, while Carolyn was still in Boulder with me. I introduced her to everyone, told them I had gotten a season ticket to watch the University of Colorado Buffaloes play football that fall, had gotten a Colorado Pass so that I could enjoy skiing the coming winter, and that I was really excited about being at Ball Aerospace in my new position. More importantly, I told them I had talked to Dave Taylor about this and that I planned to expand the role of my business unit by moving it into NASA's Exploration arena, as I thought that there were some new business opportunities available to Ball Aerospace, that were a good match between what NASA needed, and what Ball Aerospace could offer.[2] I had no idea at the time how this decision was going to play out, and things really took a strange twist for not only my business unit but, more personally, for Carolyn and me, a scant two-plus years down the road.

---

[2] If you're interested in learning more about Ball Aerospace, please refer to appendix 2.

Next Steps for NASA's Exploration Program

For the next year, things continued to move slowly in the general directions outlined earlier, at least until April 2005, when Mike Griffin became the new NASA administrator after Sean O'Keefe had decided to step down in February to become the chancellor of Louisiana State University (LSU). I should note that, at this point in time, I had been at Ball Aerospace for some seven months, so my view of what happened next is that of an informed outsider, but without the benefit of having been directly involved in the internal discussions of NASA.

Mike Griffin was a smart engineer and manager (he had seven degrees, including a PhD in aerospace engineering) who had been thinking about all this deeply while he was the head of the space department at the JHU applied physics laboratory in Laurel, Maryland, and even before then. So he came into his new position with a plan—in fact, a classic (engineering) plan for how to move things along without an infusion of new money, the nemesis of the Vision from day one. Griffin called it "Apollo on steroids," and that it was. He knew there wasn't enough money for new technology development. And he knew that he would have to forcefully manage the phase-out of the space shuttle to be able to squeeze out whatever additional monies he could from the existing NASA budget. If he didn't do that, it was highly likely that the space shuttle would continue to fly until the next accident, and he also knew that he must make sure the space shuttle flew as safely as possible, because our country, let alone NASA, could simply not afford to have another high-visibility space shuttle accident.

Part of Griffin's plan to "force" the retirement of the space shuttle in 2010 was to cancel all the contracts that supported the development of the replacement hardware that would have permitted it to fly after 2010. Of course, he knew this had to be done very carefully, lest there be another accident, so he was ever mindful of that. Accordingly, he was keen to limit the number of space shuttle flights as well, because he knew that, as some have said, the only truly safe space shuttle is one that is on the ground. He put the squeeze on NASA's science (and especially earth science) and aeronautics programs to help bring more money to the table. It should be noted that included in these changes were changes (mostly negative to the institutions involved) as to how Goddard and JPL were to be involved in the Exploration program. Among these changes were to declare null and void the agreement that Goddard and JPL had reached about which center would do what in the robotic space flight arena.

With such a plan in place, Griffin could be better assured that certain amounts of money would be available at certain points in time, and thus he would know better how to plan. As far as this book is concerned, the most relevant aspect of his plan was the development of a different type of human space flight launch vehicle with which to execute the plan. This launch vehicle, referred to as the crew launch vehicle (CLV), was named Ares, which, interestingly enough, was the ancient Greek god of warfare. Go figure. In any event, the name aside, it was a reasonable engineering type solution to the dilemma of not having enough money in the face of needing to keep things moving forward.

There were to be two forms of the Ares launch vehicle, the Ares I (CLV) and the Ares V. The Ares V was to be a huge Saturn V–style rocket that would carry cargo, not humans. The Ares I CLV was to be the human launch vehicle for taking humans to low earth orbit (LEO) and to return humans to the moon. The separation of launching humans from launching cargo, unlike the situation with the space shuttle, made the launch of humans inherently safer, since the number of moving parts, so to speak, were dramatically reduced, because the launch vehicle was much smaller. Additionally, the fact that humans would ride on top of the launch vehicle rather than alongside it (like in the case of the space shuttle), and that there was an escape tower to carry the astronauts safely away from the launch vehicle in the event of an accident, also further enhanced the likelihood that humans could survive an accident.

Our job was to focus on the Ares I CLV. It would carry something called the crew exploration vehicle (CEV), named Orion, which looks like a large Apollo capsule. And, Orion was designed to carry six humans into space (instead of three, as for Apollo). The plan at the time was to return it to earth via landing on land (parachute onto, rather than fly to the ground, like the space shuttle does), rather than in the ocean (as for Apollo). The Ares I itself relies on as little new technology as possible (to save money). Thus, the first stage is basically a space shuttle four segment solid rocket motor (SRM) augmented with an additional (fifth) segment. The upper stage consists of a tank like structure to carry the necessary fuel and oxidizer for the engine, as well as housing the engine, a modified J2 engine from the Saturn launch vehicle (i.e., Apollo) program. Immediately atop the upper stage is something called the instrument unit avionics (IUA), whose ring-like shape and terminology was also borrowed from the Saturn launch vehicle program. The IUA was to be the brains of the Ares I, as its electronics tells the Ares I launch vehicle when and where to go. This new launch vehicle was being developed via four major procurements (the first stage; the upper stage; the modified J2 engine [referred to as the J2X engine]; and the IUA). See the figure[3] on the next page that depicts the various components of the Ares I launch vehicle.

---

[3] Wikipedia contributors, "Ares I Exploded View of Major Components," accessed July 19, 2019, https://en.wikipedia.org/wiki/Ares_I.

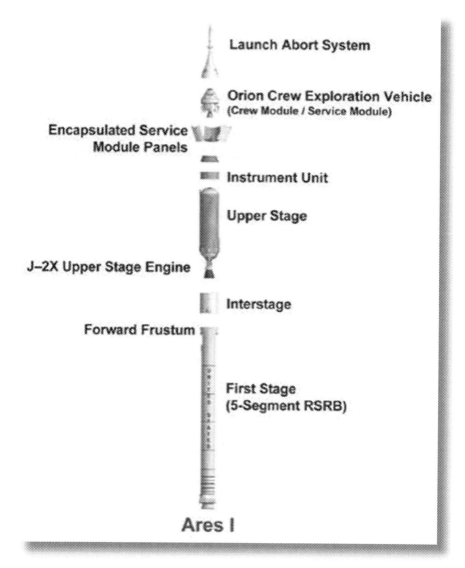

Launch Abort System

Orion Crew Exploration Vehicle
(Crew Module / Service Module)

Encapsulated Service
Module Panels

Instrument Unit

Upper Stage

J–2X Upper Stage Engine

Interstage

Forward Frustum

First Stage
(5-Segment RSRB)

Ares I

The first stage was awarded sole source (i.e., without competition) to Alliant Techsystems (ATK), and the J2X engine was awarded sole source to Pratt & Whitney Rocketdyne (PWR), both companies being the historical providers of those systems in their earlier form. The upper stage and the IUA were to be procured competitively.

The focus of the balance of this book is on the competitive procurement of the Ares I IUA.

Finding the Opportunity

Back at Ball Aerospace, we had been moving along, picking up some new work here and there in the Exploration area, but nothing that we had really been able to get our teeth solidly into yet. And then along came SpaceX and Elon Musk. For those of you who may not be aware, Elon was an independently wealthy person, having made his fortune through the development of Zip2, a provider of enterprise software and services to the media industry, which he sold to Compaq in 1999 for $307 million plus $34 million in stock options, and the development of PayPal, the online payment system, which was purchased by eBay in 2002 for $1.5 billion.[4]

Elon was also the chairman, product architect, and CEO of Tesla Motors and had led the development of the Tesla Roadster, an electric car that can go from zero to sixty miles per hour in 3.7 seconds and has a range of nearly 250 miles on a single charge.[5] Clearly, he is a true American innovator and entrepreneur. He also had a passion for space (as did I) and

---

[4] Biography.com Editors, "Elon Musk Biography," accessed April 28, 2019, https://www.biography.com/business-figure/elon-musk.

[5] Biography.com Editors, "Elon Musk Biography," accessed April 28, 2019, https://www.biography.com/business-figure/elon-musk.

firmly believed that the United States had lost its way—not just NASA but the major aerospace companies that serve NASA, in particular Boeing and Lockheed Martin, who just happened to also make launch vehicles (namely, the Delta and Atlas series). So Elon decided he would fund the establishment of a company, SpaceX, and lead the development of an alternative to these (in his view) overly expensive launch vehicles, being built by (again, in his view) overly bureaucratic companies. He was also the CEO of SpaceX, so he wore many hats.

The launch vehicle portfolio of SpaceX at the time included the Falcon 1, Falcon 1e, Falcon 9, and Falcon 9 Heavy. He was also developing something called Dragon, which was where Ball Aerospace came into the picture. But before getting into that, you need to understand the beauty of his launch vehicle approach. First, he was almost totally self-sufficient, with a manufacturing operation in Los Angeles and a test facility in McGregor, Texas. And he borrowed a page from the Russians and put all his eggs in one engine basket, the Merlin engine, which he then used over and over, depending on what he was trying to do, launch vehicle-wise. For example, the Falcon 1 first stage uses one Merlin engine, and the Falcon 1 was advertised at the time as being capable of taking approximately one thousand pounds to low earth orbit (LEO) for around $8 million (compared to his competition who charged $20–$30 million). Then, for the Falcon 9, he clustered nine of his Merlin engines onto the first stage, and designed the margins such that he had a true "engine out" capability to orbit. The Falcon 9 was advertised at the time as being capable of taking approximately 27,500 pounds to LEO for around $37 million (compared to his competition, who charged $100–$180 million). Everyone said he couldn't do it, but I met, talked to, and worked with him, and I firmly believed he would do this successfully, and in the process, beat the pants off Boeing and Lockheed Martin. Proof that he was making good progress toward providing reliable and inexpensive access to space is that on September 28, 2008, SpaceX made history when its Falcon 1, designed and manufactured from the ground up by SpaceX, became the first privately developed liquid fuel rocket to achieve Earth orbit. This was only the fourth flight of the Falcon 1, which in the launch vehicle business was a remarkable accomplishment for a totally new launch vehicle.[6] In addition, the Falcon 9 launched successfully, initially on June 4, 2010,[7] and again on December 8, 2010, that time with the Dragon capsule on board. The Dragon capsule successfully completed two orbits and then splashed down in the Pacific, where it was successfully recovered. Thus, SpaceX, with Elon Musk at the helm, became the first private company to launch a capsule into Earth orbit and recover it.[8] Since then, lower-cost launches of SpaceX's Falcon 9 have become commonplace, and the United Launch Alliance (ULA; a Boeing-Lockheed Martin joint venture) is no longer the only US provider of large launch vehicles.

Back to Ball Aerospace and its brief intersection with SpaceX—how did that happen anyway? Well, as part of my search for new business opportunities in the Exploration area, we had teamed (separately) with two other companies on their proposals to the NASA Commercial Orbital Transportation Services (COTS) request for proposal (RFP). While neither of these proposals was selected, we got the chance to develop, using Ball Aerospace internal funding, a quad redundant (having four identical electronics boxes to be able to deal with the failure of any one box), two fault tolerant (able to operate as designed after two failures) avionics architecture, which got kudos from NASA reviewers. So when the COTS winners were announced in August 2006, one of them was SpaceX, and guess what? They didn't yet have a solid avionics approach for their Dragon vehicle. One of our hotshot engineers figured this out and gave them a call. Thus, we quickly proposed to SpaceX that we could be the avionics provider for their COTS Dragon spacecraft. This led to a small contract being established between Ball Aerospace and SpaceX to support SpaceX in this area.

Not long after, I made a trip to California to sit down with Elon

---

[6] Stephen Clark, "Sweet Success at Last for Falcon 1 Rocket," accessed April 28, 2019, https://spaceflightnow.com/falcon/004/.

[7] Jeremy Kaplan, "SpaceX Falcon 9 Rocket Launches, Despite Snags and Delays," accessed April 28, 2019, https://www.foxnews.com/science/spacex-falcon-9-rocket-launches-despite-snags-and-delays.

[8] Daniel Bates, "Mission Accomplished! SpaceX Dragon Becomes the First Privately Funded Spaceship Launched into Orbit and Guided back to Earth," accessed April 28, 2019, https://www.dailymail.co.uk/sciencetech/article-1336868/SpaceX-Dragon-privately-funded-spaceship-launched-orbit.html.

and work out the details of our arrangement. While I was there, Elon personally gave me a tour of his Falcon launch vehicle factory, including showing off his brand-new automated friction stir welding machine, which was being used to weld large circular tank sections together. It was while I was there, and especially over lunch, that I got the full measure of the man, and I definitely liked what I saw. From there, based on the work that we had done on Ball Aerospace internal funding in support of the COTS proposals, we were easily able to help SpaceX get all the way through the Critical Design Review (CDR) stage successfully with the NASA reviewers once again loving our approach to the avionics system. Unfortunately, that was about the end of it, as we were never able to come to an accommodation with SpaceX on the price for Ball Aerospace to actually build this hardware/software. Elon and I had several telephone conversations about this, but we simply weren't able to go where he wanted us to go pricewise. So we parted ways amicably, and I wished him well. I still do. While it wasn't to be with SpaceX, this overall set of experiences involving the COTS proposals and SpaceX helped build our confidence relative to pursuing an exciting new opportunity that we had identified in the Exploration area (i.e., the Ares I Instrument Unit Avionics [IUA] procurement being run out of NASA's Marshall Space Flight Center in Huntsville, Alabama).

Not long after I arrived at Ball Aerospace in 2004, Dave Taylor decided to reorganize various aspects of the company via an activity that he called rebalancing. One aspect of this was to create a corporate business development (CBD) unit that was charged with identifying new business opportunities in a way that made them known and available to all aspects of the business, independent of the specific objectives of any single business unit. As part of this process, I gave up some of my new business folks, one whose name was Dave Murrow (no relation to Dave Taylor or Dave Hoover, just another Dave) to this new unit. Dave Murrow was a bright, ambitious fellow with a solid background in the technical and new-business arenas. His new job was to help uncover new opportunities in a variety of areas, most notably including Exploration. So Dave and I talked a lot, and he went out and shook the trees as hard as he could to see what fell out of them.

Dave Murrow had figured out that an opportunity of possible

interest was in development at NASA Marshall, in Huntsville, Alabama. Discussions between him, other Ball Aerospace personnel, and Marshall began in 2005. Frequent travel to Marshall to talk to them and their partners allowed Dave to get a better feel for what might be coming up. Things heated up a bit in early 2006, when Dave and others met with several of the key personnel at Marshall, including Steve Cook, then director of the Exploration Launch Projects Office.

Along the way, Dave started meeting with me monthly to discuss what he and his team were coming up with. At first (these initial discussions predated our involvement in the NASA COTS activity) I was skeptical, as was my senior staff. Launch vehicles? What could Ball Aerospace offer, and how could we possibly be competitive? But as we learned more, it began to sound more interesting, and I agreed to provide some funding and other support to help move things along.

The next real turning point was in September 2006 when Dave arranged for Steve Cook and his Marshall team to visit Ball Aerospace. I got personally involved in this visit, as did Dave Taylor. At the conclusion of these discussions and demonstrations, we all went to dinner together in an effort to get to know one another better. I brought Carolyn along, and, as it turned out, she knew some of the folks from her days at NASA Headquarters working in the human space flight arena. So that was a real icebreaker, and we ended up having some really good discussions. Marshall was working hard to keep the Ares I IUA procurement separate from the Ares I Upper Stage procurement because they had visions of Wernher von Braun, the father of the Saturn V Apollo launch vehicle, dancing in their head. As I described earlier, the Ares I IUA terminology came from the Saturn V launch vehicle, which had a structural ring the diameter of the launch vehicle, with all the avionics (electronics) attached to it (during the time period of this procurement you could actually see one of the leftover Saturn IUA units at the US Space and Rocket Center in Huntsville, Alabama, the home of Marshall). This was the ring that the Saturn V engines got all their instructions from, as the human race headed out to the moon for the first time ever. But the main point to be noted here is that von Braun set things up so that the Saturn V Instrument Unit Avionics (IUA) was designed by Marshall engineers (in-house) and then built by IBM. (An interesting aside here: when we eventually moved to

Huntsville with our team, we coincidentally occupied the same building at 200 Sparkman Drive, which IBM had occupied when they were building the Saturn V IUA's for Marshall.)

Marshall wanted to replicate that experience. They wanted to design the Ares I IUA and then have an aerospace industry company build it. They felt this was the shortest path to reestablishing their credibility within NASA as a leader in the launch vehicle arena, much as they had been in the von Braun days.[9] Well, given Ball Aerospace's storied fifty-year history of innovation and creativity in support of NASA, our demonstrated ability to let NASA lead the way when they wanted to, and our recent successful experience with launch vehicle avionics, especially with SpaceX, we knew at this moment in time that this was a gold ring we should try to grab. Thus, with the full cooperation of Dave Murrow's boss, the vice president, CBD, Art Morrissey, I assigned Dave Murrow as the capture manager for this opportunity, and my business unit took over the pursuit from CBD. So the ball was not only rolling but also picking up speed.

---

[9] If you're interested in learning more about either Huntsville, Alabama, or the NASA Marshall Space Flight Center, please refer to appendix 1.

The Ares I IUA: The Job

So, what did NASA want industry to do? As I noted earlier, Marshall's dream was that they would design the IUA, and industry would build it. But that was initially just a dream. First, they had to overcome NASA Johnson Space Center in Houston, Texas, to whom Marshall reported, and who was hell bent on adopting a common avionics architecture between their crew exploration vehicle (CEV) and Marshall's Ares I IUA. This precipitated a nasty set of arguments among NASA Headquarters, Johnson, and Marshall, and finally a decision was made to have an independent review of the issue done by the agency-level NASA Safety and Engineering Center (NESC), located at NASA Langley Research Center in Hampton, Virginia. The NESC was established as a partial response to the Columbia accident and was intended to help ensure the safety and engineering excellence of NASA's programs and institutions. The NESC was set up to accomplish this by performing value-added independent testing, analysis, and assessments of NASA's high-risk projects.[10]

As you might imagine, this review was not completed overnight, and

---

[10] NASA, "NASA Engineering & Safety Center Mission," accessed April 25, 2019, https://www.nasa.gov/nesc.

we at Ball Aerospace were watching it like a hawk, because if Johnson had its way, Honeywell (a subcontractor to Lockheed Martin, who had won the CEV procurement) would most likely merge the requirements for controlling the Ares I with those for the CEV. In that manner, the whole CLV/CEV system would be controlled by one set of avionics located in the CEV, and the IUA contract competition would never occur. But as it turned out, the NESC recommended that the Ares I avionics (IUA) be separate and distinct from the CEV avionics for many good and valid reasons, not the least of which was to provide another level of independence with respect to helping ensure the safety of the humans that would be carried into space by this combined CLV/CEV system, and also to permit the Ares I IUA design to more naturally flow to the future Ares V (cargo) launch vehicle, which would not have a CEV with its avionics, on top of it. This was a major decision, which not only ensured that there continued to be a potential role for Ball Aerospace in this procurement but also signaled that Marshall thinking was beginning to get some additional traction within the human space flight community, which, in the post-Apollo era, had been dominated by Johnson.

The next hurtle for Marshall was figuring out how to get the design job done. Their engineering talents had been significantly weakened from the days of the Saturn V development, and it didn't appear to me that they could do the job by themselves. Apparently, Marshall figured that out and surrounded themselves with support contractors such as Draper Labs and others drawn from the extensive engineering support services community of Huntsville, Alabama. Additionally, we heard rumblings that the Ares I IUA RFP would have extensive provisions for Indefinite Delivery, Indefinite Quantity (IDIQ) work by the winner of the Ares I IUA procurement. This signaled to us that they would be looking to the IUA contractor for engineering assistance as well. In thinking about this, we saw a potential new role for ourselves—that of an honest broker. As such, we could work side by side with Marshall, thus forming a badgeless team, and be able to provide them with the benefit of our extensive experience, such that the design was the best possible, while allowing Marshall to "own" it, as they wanted to do. We later confirmed that this was, in fact, something that Marshall wanted very badly. So we strove hard to give it to them.

Another related issue was Marshall's ability to handle the many major procurements that were required to make the Ares I launch vehicle go forward on schedule. Marshall had had a recent history of being quite lax in this area, but Mike Griffin, the NASA administrator, made it very clear to Dave King, the Marshall center director, that Marshall must meet their procurement schedule, or not only would heads roll, but the procurements would be given to Johnson, something that would have forever relegated Marshall to a minor role in the return of humans to the Moon, and would have called into question the very reason for Marshall's continued existence as a NASA field center. So the stakes were very high here.

All of this made perfect sense to me, as I had had an earlier discussion with Mike Griffin at one of the space shuttle launches from the NASA Kennedy Space Center in Florida, and asked him what he hoped to accomplish before his tenure was over (which was likely to happen at the end of the Bush Administration in early 2009). He told me he wanted to accomplish three things:

1) Retire the space shuttle by 2010
2) Complete the International Space Station (ISS) so that NASA's International Agreements could be met
3) Get enough traction under the Vision for Space Exploration such that it could be sustained after his departure, including sustaining it in the face of a likely party change in the White House and possibly the Congress as well.

Note that he said nothing about NASA's science or aeronautics programs. Clearly, this was a man with a very specific, focused agenda, and that agenda had a largely unmovable deadline associated with it. To Marshall's credit, they did maintain their schedule, undoubtedly because there was so very much focus on it by NASA Headquarters, Johnson, and certainly Dave King and his senior staff at Marshall.

Thus, the Ares I IUA RFP was released almost exactly on schedule in early June 2007. Now, we actually knew what NASA wanted, as opposed to trying to piece it together from meetings and rumors. As expected, Marshall wanted to do the design, and they wanted their contractor to build the IUA. This included the procurement of avionics boxes, physical system integration, multiple-unit production, and test of twenty-one quad redundant, two fault-tolerant flight articles, with most of the work to be performed in Huntsville, and at the NASA Michoud Assembly Facility located in New Orleans. The contract was to be cost reimbursable. Our proposal was due in late July, and the anticipated contract award date was November 2007 (later changed to December 2007). The work was to be split into five contract line items (CLINs):

1) CLIN 1—Design, development, test, and evaluation (DDT&E) leading to flights of a development unit, a qualification unit, and a full-up flight unit, all during the first six years of the contract (Contract Year [CY] 1–CY6)
2) CLIN 2—DDT&E support via IDIQ (CY1–CY6)
3) CLIN 3—Baseline Production at a rate of two quad redundant, two fault-tolerant units per year (CY4–CY9)
4) CLIN 4—Production Support via IDIQ (CY4–CY9)
5) CLIN 5—Optional Production at a rate of up to four additional quad-redundant, two fault-tolerant units per year (CY4–CY9)

While it may not be immediately apparent, this was a pretty big deal. From a cold start, the expectation was that a brand-new design would be developed and certified via flight test early enough to commit to production of flight articles in the fourth year of the contract after which as many as an additional eighteen quad redundant, two fault-tolerant flight articles (six firm and twelve optional) might be built by the end of the ninth year of the contract. And the articles being designed, tested, and built were very complex.

Breaking it down into its simplest form, each deliverable consisted of four identical electronic units. Each unit needed to have at least a power subsystem; a guidance navigation and control subsystem; a command and data-handling subsystem; an RF system for communicating back to Earth from along the flight path; as well as several complex sets of test equipment to run all this hardware through its paces and check it out. So that's four major subsystems per unit, times four (redundant) units per deliverable, times twenty-one deliverables, which is a minimum of 336 subsystems. The actual number is much, much larger than that, once you add in the many other related, but smaller, subsystems and components, including interface boxes, sensors, cabling, and so on. As further evidence of the magnitude of this job, when this contract was awarded, its stated full value was just under $800 million, not an excessively huge contract value, but certainly not a small one either. It was very clear that whomever won this job would have a core piece of significant work for many years to come, plus a major role for a very long time in the nation's plans to return humans to the moon and go on to Mars. The gold ring was getting shinier.

The Competition

As you can probably imagine, for good and valid strategic reasons, most companies don't exactly climb up on a box and announce to the world that they are competing for a particular job, especially if it's a big one, which the Ares I IUA was. So it falls to an individual company's business development office and, as appropriate, their Washington, DC, office, to try to figure this out from hallway talk, rumors, watching the comings and goings at NASA Headquarters and other locations, and listening to your known competition whenever they talk at conferences and such to see what they are saying as ways to try to piece that part of the story together. Eventually, you may be able to get some sense of who is doing what, but even then you have to keep listening carefully, because, as frequently happens, as the time for the RFP to come out gets closer, some companies decide to team with other companies, and some decide to drop out, and once again, you don't know for sure who is doing what. This may come as somewhat of a surprise to some folks, especially those who have never worked in industry, but this part of the puzzle, and most things related to it, or strictly company confidential, and theirs to know, and everyone else's to try to figure out. Of course, it goes without saying that knowing who your competition is, is absolutely critical to knowing the best strategy to use to win the job.

So who was Ball Aerospace's competition for this job? Well, for reasons that will become clear in the next chapter, in the end, there was a lot of competition. Four other companies bid this job, which for a procurement of this size and type is a lot more than would usually be expected.

The four other companies were as follows:

1) BAE Systems,[11] sometimes referred to as British Aerospace. BAE Systems is a global company engaged in the development, delivery, and support of advanced defense and aerospace systems in the air, on land, and at sea. They had major operations across five continents, with customers and partners in more than one hundred countries. In 2007, they employed almost one hundred thousand people, and their annual sales exceeded £15.7 ($20.4) billion. They already had an office in Huntsville as well. They did not have any significant contracts with NASA in the human space flight arena.

2) Raytheon,[12] a global defense and aerospace systems supplier, is focused in four core defense markets, mission systems integration, homeland security, and information operations/information assurance areas, as well as technology development and mission assurance. They consider themselves to be a global leader in technology-driven solutions. In 2007, they employed more than seventy thousand employees worldwide, and had more than $23 billion in sales. They did not have an office in Huntsville, nor did they establish one there. They also did not have any significant contracts with NASA in the human space flight arena.

[11] BAE Systems, "Company Description," accessed February 3, 2009, https://www.baesystems.com.
[12] Raytheon, "Company Description," accessed February 3, 2009, https://www.raytheon.com.

3) Honeywell,[13] teamed with Lockheed-Martin.[14] Honeywell International is a diversified technology and manufacturing leader, serving customers worldwide with aerospace products and services; control technologies for buildings, homes, and industry; automotive products; turbochargers; and specialty materials. In the aerospace area they considered themselves to be the world's premier supplier of aircraft engines and systems, avionics, and other products and services for airliners, businesses, and general aviation aircraft, military aircraft, and spacecraft. In 2007 they had almost 130,000 employees in more than one hundred countries and had sales of almost $35 billion. Honeywell's teammate, Lockheed-Martin, is a well-known major aerospace company as well, with 140,000 employees, and sales of about $42 billion in 2007. Honeywell did not have an office in Huntsville, nor did they establish one there.

4) Boeing[15] is one of the world's leading aerospace companies and the largest manufacturer of commercial jetliners and military aircraft combined. Additionally, Boeing designs and manufactures rotorcraft, electronic and defense systems, missiles, satellites, launch vehicles, and advanced information and communication systems. As a major service provider to NASA, Boeing operated the space shuttle and the International Space Station. In 2007, Boeing employed more than 160,000 people across the United States and in seventy countries and had sales of approximately $61 billion. They had had an office in Huntsville for more than thirty years.

This should give you some sense of who and what Ball Aerospace was up against. Ball Aerospace had annual sales in this timeframe of around $700 million, with about three thousand employees. Thus, Ball

---

[13] Honeywell, "Company Description," accessed February 3, 2009, https://www.honeywell.com.

[14] Lockheed Martin, "Company Description," accessed February 3, 2009, https://www.lockheedmartin.com.

[15] Boeing, "Company Description," accessed February 3, 2009, https://www.boeing.com.

Aerospace is not in the same league as the other four companies. These four companies, five if you include Lockheed Martin, are all clearly the big boys of the aerospace industry, which meant that however you cut it, Ball Aerospace would have to propose something really creative, innovative, and affordable to be able to win this job. That became exactly our goal.

Like Ball Aerospace, BAE Systems and Raytheon saw this as a potential new market for their capabilities. So, together with Ball Aerospace, they were also new kids on this particular block.

As mentioned earlier, Honeywell had teamed as a subcontractor to Lockheed Martin, the winner of the crew exploration vehicle (CEV), or Orion, procurement. Honeywell was to be the avionics provider for Orion. For the Ares I IUA procurement, the teaming arrangement was reversed, given that the Ares I IUA was strictly an avionics job. While Lockheed Martin won the Orion procurement, with Honeywell as their teammate, they were both relatively new players in the human space flight arena. Thus, this team clearly saw a chance to extend their reach from Orion into Ares.

And then there was Boeing. And Boeing being, well, Boeing, they were clearly the incumbent in this field (if there is such a thing for a brand-new launch vehicle development). It is always extremely hard to knock off a deeply entrenched, long-established incumbent contractor in any area that one might strive to enter as a new player, and this situation was no different. For this procurement, the lay of the land was clearly that there was Boeing, and then there were four other companies, each trying in their own unique way to overcome Boeing. And we all knew that would be a formidable task. But NASA was saying things that gave the four of us the much-needed encouragement and hope to try to do just that.

What Was NASA Saying?

Typically, the "black curtain"—whereby NASA stops communicating with its prospective contractors—falls around the time of the RFP release, but sometimes it happens earlier. It is always in NASA's best interests to keep their doors open as long as possible before pulling this curtain down, so that the period before the curtain falls is a time that contractors must use very effectively to get as much additional understanding out of NASA and to learn as much as they can about what NASA really wants.

So what were we hearing from NASA during this time period? It was indeed saying something to all the potential new players that was very important and, in fact, critical to whether or not anybody but Boeing would have even half a chance at winning this competition. Unfortunately, certainly in the early days well in advance of the RFP release, NASA's story wasn't always internally self-consistent, especially between Marshall and NASA Headquarters. While their story got much more consistent with time, this inconsistency was a concern.

The way this works is that the government—NASA, in this case—leads, and industry follows. Sometimes industry is able to convince NASA to do things differently, but not very often. Thus, it becomes industry's job to respond to NASA's requirements as well as they can, while delivering the required products within specification, on a particular schedule, and for a cost that fits NASA's budget. Of course, industry never really knows how much NASA is budgeting for the task in question, so that becomes another aspect of what industry has to deal with to be able to submit a winning bid.

First, as noted earlier, Marshall wanted desperately to keep the Ares I IUA procurement separate from the Ares I Upper Stage procurement. Had they not been able to do that, I believe there would, practically speaking, have been no opportunity for Ball Aerospace, or BAE Systems and Raytheon, to have participated in the procurement, given that none of these three companies have sufficient (or any, in some cases) launch vehicle capability per se, to have been able to have bid the nonavionics portions of a single Upper Stage procurement. Perhaps some of us could have teamed with someone else, but that would have been the extent of it. As far as Honeywell and Lockheed Martin goes, I suspect they could have continued to play ball, since Lockheed Martin is a long-time launch vehicle provider via their Atlas and Titan series of rockets, but Lockheed Martin would have undoubtedly proposed as the prime contractor, with Honeywell the subcontractor, much as had been the case for Orion. But Marshall got their way, and the two procurements were indeed kept separate.

The next key aspect of this story is that Marshall, up and down the management chain, said that they wanted two separate contractors for the Upper Stage and the IUA. Their principal logic for this was that they were afraid that if the same contractor won both jobs, than that contractor would have been very well motivated to propose, after they won, to merge various portions of the job, the management thereof being one example, but technical aspects of the job as well, such that Marshall would have eventually lost its identify as the owner of the Ares I IUA design, a key objective of Marshall throughout this procurement.

Marshall wanted new blood in the program. This was said virtually everywhere we went in the community, and it made perfect sense, since having new blood meant having another voice out there supporting and advocating the Exploration program to the Congress. Additionally, since Marshall was intent on developing and owning the design for the Ares I

IUA, they felt they needed someone who was new to the game and would be more likely to be responsive to Marshall. Someone who would be willing to step back into the shadows whenever Marshall wanted to bask in the limelight of their anticipated success. We heard over and over that Marshall absolutely did not want to have to deal with Boeing on this job, as Marshall was convinced that Boeing would do what Boeing wanted to do, not what Marshall wanted Boeing to do.

We also heard, mainly from Steve Cook, the head of the Exploration Launch Vehicles Project Office at Marshall, that NASA wanted to learn from the aircraft industry. He said many times, publicly and in meetings, that he did not want another space shuttle avionics approach, which did not easily permit upgrades of capabilities, and had kept NASA locked into specific suppliers for thirty to forty years. Thus, NASA wanted to see innovative and highly efficient approaches migrate from the aircraft industry to the space industry, at least as far as designing, building, and upgrading avionics went.

So these four key things were what we heard NASA telling us and what convinced us that we actually had a chance of winning this procurement:

1) Keep the Ares I Upper Stage and IUA procurements separate.
2) Select two different contractors for the two jobs.
3) Select new blood for the IUA.
4) Demonstrate that you can migrate innovative aircraft industry approaches into the Ares I IUA effort.

What NASA was saying directly to us, and others, convinced us that we had at least a fighting chance to win this procurement. Additionally, on the surface of things, it seemed that our real competition was the non-Boeing players. However, Boeing was always in the back of our minds, such that we eventually decided to approach them about them teaming with us as a subcontractor to us, something that seemed pretty brash on our part, which it was, but more on that later.

The Decision Maker

The die had been cast quite a bit earlier, that the procurement decision for the Ares I IUA, and other similar Exploration procurements, including the Ares I Upper Stage, would be made in Washington, DC, at NASA Headquarters. While it would have been unusual in the robotic side of NASA to have major procurement decisions made at NASA Headquarters, rather than at the field center, this was not unusual in the human space flight side of NASA, as NASA Headquarters historically had had all it could do to keep the NASA Johnson Space Center, the lead center for implementing the Exploration program, from becoming the Johnson Space *Agency*, and this was one small way to help defend against that.

So who would this person most likely be? Normally it would have been the associate administrator for the Exploration Systems Mission Directorate, who at the time was Scott (Doc) Horowitz. But Doc, previously an astronaut, had come most recently from industry (ATK-Thiokol) and had to recuse himself from such decisions. Thus, the task fell to his deputy, Doug Cooke. And that's the way it appeared to be until August 2007, when, after Doc Horowitz decided to return to industry in October, the announcement was made that he would be replaced by then NASA Stennis Space Center director, Rich Gilbrech. While we were never sure of this until the very end, possibly for continuity reasons, Doug Cooke ultimately remained the decision maker for the Ares I IUA.

Doug Cooke[16] had more than thirty-five years of experience at NASA in the space shuttle, space station, and Exploration programs. He was assigned significant responsibilities during critical periods of each of these, including top management positions in all three programs. He certainly seemed like the right person for the job. I only got to meet with him once, since he decided to pull his black curtain down much earlier than anyone else involved with the Ares I IUA procurement. That meeting was on December 1, 2006. Carol Lane, the head of Ball Aerospace's Washington office was with me and had set up this meeting.

The main surprise for me was the degree to which he seemed disconnected from the Ares I IUA procurement. He seemed much more interested in talking to me about an upcoming Exploration conference that he was to be involved in. Was it simply too early for Doug to be significantly involved, or, as I had seen happen many times during my eighteen years at NASA Headquarters, was the NASA field center in question, Marshall in this case, simply keeping him in the dark? While I didn't know the answer, the nonspecific nature of our discussion led me to think that Steve Cook, no relation, and the director of the Exploration Launch Vehicle Project Office at Marshall, might, in fact, be the main person driving this procurement.

Nonetheless, Doug and I talked about some things he seemed to have an opinion on, such as teaming, the idea of a contract that migrated from cost plus to fixed price once the design phase was complete, among other things. And, of course, I did get an ample opportunity to tell him why I thought Ball Aerospace was the right contractor to do this job.

So, on that note, things started to get much more serious very quickly.

---

[16] NASA, "Douglas R. Cooke," accessed February 3, 2009, https://www.nasa.gov/exploration/about/cooke_bio.html.

PART 2

Before the Storm

The Ares I IUA: A Must-Win

There is nothing particularly magical about a pursuit being designated a must-win by a company. It is fairly standard terminology in the industry and simply designates that a particular activity is crucial and has significant implications (up or down) for the future of the business. The business will not collapse if a must-win isn't won, but there are usually significant negative implications for the business when that happens. Being designated a must-win also means that that particular pursuit commands greater resources than normal so as to improve the probability of a win. A must-win does not mean other pursuits are not important.

As mentioned earlier, I had been meeting monthly with Dave Murrow and the pursuit team for quite some time. Since things were definitely beginning to get more serious, I decided to ratchet the heat up a bit by elevating these meetings to the level of Dave Taylor, the CEO and president of Ball Aerospace, and my boss. These regular meetings with Dave Taylor started in November 2006.

Once Dave Taylor got involved, it was much easier to get support from other sectors of the company for efforts that were beyond the ability of my business unit to handle totally by ourselves. Thus, Dave and I were able to work together to deal with a number of such issues, most notably the idea of establishing an office in Huntsville, Alabama, as a way to get closer to the customer, and to demonstrate further commitment to the job by being on-site in Huntsville. Dave easily bought into this, as the benefits to the pursuit of doing this were obvious. By the time of our second monthly meeting with Dave in December 2006, we had made considerable progress toward identifying possible locations for such an office. We quickly concluded that locating this office in space available at 200 Sparkman Drive in Cummings Research Park directly across the street from the University of Alabama, Huntsville (UAH) was the thing to do. The lease was quickly signed, and we were then the proud tenants of an empty office space (about five thousand square feet) right in Marshall's backyard. As noted earlier, this was some of the same space that IBM had occupied when they were building the Saturn V Instrument Unit for Marshall back in the 1960s. Once we discovered this, we considered that a good sign for the future. I later found out from Dave King, the Marshall center director, that he drove by this office every day as he took one of his children to childcare. Another good sign.

The lease was effective on February 1, 2007, so our plan was to open the office on February 20 with a gala ribbon cutting ceremony to include the mayor of Huntsville, Loretta Spencer. Our initial plan had been to move a couple of folks to Huntsville to staff this office, so the not quite three weeks between being able to move in, and having the ribbon-cutting ceremony, were spent getting phone lines installed, internet connections established, furniture moved in, and getting those first couple of folks on the ground in Huntsville.

So, on February 19, 2007, several of us, including Dave Taylor and me, headed to Huntsville for the opening of the new Ball Aerospace office the next day. The event went well, and we had a good turnout for the event, complete with snacks, beverages, and such. The highlight was having the mayor participate with Dave in the ribbon cutting. They talked quite a bit, and soon became "best buds." The mayor even told Dave about the wonderful Huntsville Botanical Garden, where she volunteered a couple of hours a week, doing weeding and things like that (she was nearly seventy). She invited Dave to come see the gardens with her, an offer that we later turned into another significant public event, thanks to my wife, Carolyn, in her role as our community liaison.

There had been reporters and photographers present for the office opening, and an article quickly appeared in the *Huntsville Times* newspaper. Now Ball Aerospace was officially in Huntsville, but with only a small physical footprint, and an even smaller staff. But that was about to change.

In the February monthly meeting with Dave Taylor, two especially key decisions were made. First, a decision was made by Dave to elevate the Ares I IUA to the status of a must-win. This changed things yet again. Another attribute of a must-win, not mentioned earlier, is that once a pursuit is declared a must-win, it is no longer strictly a single business unit effort; it becomes a company-wide effort. Accordingly, meetings are held, not just with the VP of the business unit (me), or the CEO of the company (Dave,) but with Dave and his entire senior staff. In this way issues can be raised and dealt with much more effectively and expeditiously.

The second key thing was that we got the endorsement of an idea to engage some expert outsiders in a review of our strategy and plans for moving forward with the pursuit. This group consisted of

- John Goyak and Roger Roberts of Goyak and Associates (John was ex–Lockheed Martin, and Roger was ex-Boeing);
- Jim Madewell, president of Artic Slope Research Corporation (Jim had had a long history with NASA human space flight programs, both inside and outside NASA);
- Parker Counts, ex-NASA Marshall (a former key player in Marshall's portion of the space shuttle program, more specifically as manager of the External Tank, or ET, as it was known); and
- Carol Lane, who heads up Ball Aerospace's Washington, DC, office.

This senior strategy review group, as we called them, met for a full day on March 7, 2007, and reviewed in depth our strategy, plans, and ideas for how we were going to win this job. I was present at the beginning of the meeting, and then received the out brief at the end of the day, after which we out briefed Dave Taylor. A key discussion took place with me when I met with the group. They had concluded that the most important thing that needed to be done to have any chance of this endeavor being successful was to have a senior, full-time leader of the pursuit to not only lead the way and break down the institutional barriers but to be recognized by the customer as someone who was very credible in the industry, someone they knew, respected, and trusted.

When I asked, "Who do you have in mind?" Jim Madewell replied loudly, "You!" (meaning me), which was the absolute furthest thing from my mind. You could have knocked me over with a feather. Not too surprisingly, I initially protested, but this idea definitely had legs, and I knew right away that I was going to have to deal with it.

Putting myself aside for a moment, we'd had serious internal discussions about doing something like this ourselves. And we'd tried but were unable to close an HR deal with any of the really senior folks around the industry that we had approached, so this idea had gone nowhere within Ball Aerospace up to this point.

When we briefed Dave Taylor on all their findings and recommendations, which also included the idea of moving the entire pursuit team to Huntsville, the one involving me got considerable airtime, but Dave simply said he would have to think about that.

I went home that night and asked my wife, Carolyn, if she had a nice dinner planned with some wine, as we needed to talk. And talk we did, until one in the morning. But, after a bottle of wine, we came to a reasonable accommodation as to how we might be able to do this, which, interestingly enough, included Carolyn's participation. I knew we needed to become quickly integrated into the Huntsville business community, so we became a known commodity, both around town and at Marshall. I asked Carolyn if she would come out of retirement to help Ball Aerospace win this job. She had retired from NASA in 2005 after thirty-five years of working in the congressional, public affairs, and outreach areas, all having to do with human space flight. She had even been a charter member of the space station task force that had orchestrated the initiation of the space station program back in 1984. I was certain that she was the right person to do what needed to be done in Huntsville. And after more wine, she agreed.

Now all I had to do was to sell Dave Taylor on my general approach to being able to take on this new assignment, as well as the need for

Carolyn's participation. As far as Carolyn being directly involved, Dave was persuaded by my rationale, and she soon came on with Ball Aerospace as a part-time community liaison reporting to Sarah Sloan, the head of Ball Aerospace's communications group. Carolyn was then assigned to the new Ball Aerospace office in Huntsville.

We also needed to figure out what to do with my Boulder business unit when I went to Huntsville. It took Dave a while to get this one sorted out, but things were pretty much in place, but not well known, by the time of our next must-win meeting in mid-March. Getting an endorsement of the plan in that meeting led to me holding an all-hands meeting with the Ares I IUA pursuit team, followed by Dave holding an all-hands meeting with my entire Civil Space Systems Business Unit, and the Operational Space Business Unit, both on March 19, 2007. This was because Dave had decided to merge these two units under Cary Ludtke, then the vice president and general manager of the Operational Space Business Unit. The combined organization was to be called Civil and Operational Space (COS). This was a good move for the company, as it allowed me to almost immediately go to Huntsville to lead the Ares effort, yet it quickly put my old business unit in good hands.

With all this done, Dave Taylor and I headed to Washington, DC, to meet with Rex Geveden, then the number-three person at NASA Headquarters, and formerly the Marshall deputy center director. I even managed to have my administrative assistant, a wonderful lady named Marlene Kruise, rush new business cards through the process for me. This enabled me to be able to hand them to Rex, and others at NASA Headquarters, the next day, with my new title, Vice President, Exploration Systems, and my new location, Huntsville, Alabama, on my new business cards. Things were really moving along now.

The Move to Huntsville

Another important aspect of my becoming the senior executive for the Ares I IUA pursuit was that we felt that, by me leading by example (my actually moving to Huntsville), it would make it easier for the Boulder folks who were supporting the pursuit to do likewise. And I had a number of people tell me that it made a difference to them, that my being willing to make this move, rather than just telling them to move, was a good thing indeed.

So the move was a big focus of my March 19 all-hands meeting with the Ares I IUA pursuit team. I also went over all that the company was doing to get behind this pursuit as a true must-win, announced a handful of key folks that I had lined up behind the scenes to go to Huntsville with me, and filled them in on some of the next steps that we would be taking once we were firmly in place in Huntsville. I also spent a fair amount of time explaining what my role would be in Huntsville. For example, I would not be the capture manager. That would still be Dave Murrow (who had also agreed to move to Huntsville). Nor would I be the program manager suggested in our proposal for the execution phase. We were still working on that one. Rather, I would be the executive leader of the effort, the senior customer interface both in Huntsville and in Washington, DC. I would still report directly to Dave Taylor, and, perhaps most importantly, I would facilitate the provision of critical resources by my colleague VPs back in Boulder. I also shared all the customer meetings that were being set up for me, and in some cases for both Dave Taylor and me.

I talked about Carolyn and my thoughts on the sort of things that we would be doing in Huntsville to quickly integrate Ball Aerospace into the business community. This included sponsoring events, local congressional office visits, regular press releases, attending local business community functions, participating in the Rocket City BBQ, and sponsoring a special showing of the movie *In the Shadow of the Moon* for the business community, including Marshall. This movie, directed by Ron Howard, took a totally different look at how America had gotten to the moon, one told almost solely through the eyes of the astronauts who had gone there. It had been a film that Ball Aerospace had sponsored at the Boulder International Film Festival (BIFF) in February 2007, which had gone over well, having won top honors at BIFF, and which had also gotten awards at the Sundance Film Festival prior to being shown at BIFF. Figuring out how best to make this showing in Huntsville happen would elude us for a while, but eventually we did show it in the fall of 2007.

While this was all perceived as good, it came time to talk about exactly who we needed to move to Huntsville. At that time, we were looking at an initial pre-RFP release team of about thirty-five people, about two-thirds who we needed to be on-site full time in Huntsville. As you can imagine, some folks had family issues that kept them from easily agreeing to make this move, but we got a very positive response and ended up with an excellent team in Huntsville, which, after the RFP release, grew to about sixty-five people, including our teammates.

During March 2007, I made two trips to Huntsville and one to Washington, DC, both to meet with customers and to continue to shake things out in Huntsville relative to the impending move. Then in early April, Carolyn and I went to Huntsville together to meet with various folks in the business community relative to executing our plans for getting integrated. One of these meetings was with Rick Davis from the Huntsville Chamber of Commerce. His job was to help new businesses get settled into the community. He also was specifically responsible for doing likewise for the folks who were located in Cummings Research

Park, as we were. Listening to his ideas, descriptions of opportunities, and recitation of various upcoming events was like drinking from a fire hose. But that meeting laid the specific groundwork for most of what we did in Huntsville, and Rick was certainly instrumental, then and later as well, in our success in getting quickly integrated.

The team went down in early April to get settled into their new surroundings. I wanted them up and running before the planned April 17 draft RFP release. We were bursting at the seams at the Ball Aerospace office as we had never planned to have twenty to twenty-five people there, and quite frankly, there really wasn't adequate room. We had three or four people in each of the available one- to two-person offices, and everybody else hanging out on tables in our lobby, so we searched for additional space. When we moved into our new sixteen thousand square feet of additional space on Williams Avenue in May, it was none too soon.

Other than the early April trip just as the team began to arrive, Carolyn and I stayed in Boulder, mainly so I could attend the National Space Symposium (NSS) in Colorado Springs before actually making the move to Huntsville. I met with lots of prospective teammates at the NSS, but there was one key interaction that I'd like to share.

We had been trying to figure out what to do about Boeing, since we knew they were the longstanding incumbent in human space flight, and in Huntsville. So Dave Taylor and I decided to meet with Brewster Shaw at Colorado Springs. Brewster was the head of Boeing's NASA activities, and I knew him from when we had both been at NASA. He had been an astronaut, having flown on the space shuttle three times (STS-9, STS-61B, and STS-28). When I first met him, we were both at NASA Headquarters. He was the deputy program manager for the space shuttle program, but stationed at NASA Kennedy, and I was the deputy associate administrator (programs) for the Mission to Planet Earth (MTPE) enterprise. When earth science payloads flew on the shuttle, which was frequently during the '90s, I was usually the senior payload person involved in the launch process, which is how Brewster and I got to know each other.

Brewster had moved to the private sector, as many of us did, and he was doing well at Boeing. Dave and I met with him to talk about teaming on the Ares I IUA pursuit. At first, Brewster didn't get it. We were proposing that Boeing team (as a subcontractor) with Ball Aerospace (as the prime contractor), and I think he initially thought we meant that we wanted to do this the other way around. Our logic for proposing it the way we did was that if NASA really wanted new blood, like they were saying they did, then Boeing was DOA for this procurement. But if they got onboard with us as the prime, they could still get some work out of this, and we would have the much-needed depth that we were looking for in a teammate. So, nothing ventured, nothing gained, as they say.

In any event, when Brewster finally got in line with the discussion, he just smiled, as if to say, who do you guys think you are? This is Boeing that you're talking to, don't you know? So we shook hands and amicably parted ways and that was that, although we did approach Boeing another time in Huntsville about teaming, and that story is coming up.

Coming back from Colorado Springs, Carolyn and I seriously began to get ready for the move. By the time we got everything packed up, we had a substantial mix of checked and carry-on bags, sixteen in all, to take on the plane with us. So on April 17, we headed out to the Denver airport and were off on a totally new adventure.

Arriving in Huntsville, imagine my surprise when I discovered that the long-term leased rental car that Ball Aerospace had reserved for us was a Ford Mustang. Carolyn, me, and sixteen bags in a Mustang? Right! But we made it to the Embassy Suites in Huntsville and checked into room 835, a nice two-room suite overlooking Big Spring International Park and downtown Huntsville (see below). This was to be our home for the next eight months.

Getting Integrated into the Community

As the community liaison on-site in Huntsville, getting Ball Aerospace quickly integrated into Huntsville was Carolyn's job, with help (and funding) as appropriate from Boulder. As noted earlier, we arrived in Huntsville for the duration on April 17, and Carolyn had to hit the ground running with both feet. She had only been officially onboard Ball Aerospace for about a week, had just gotten a laptop and a BlackBerry, neither which were set up, so it was a challenge for her, to say the least. The big driver was that we had planned a suppliers conference for April 27, which was only ten days away. While she had been involved in the planning for this a bit before arriving in Huntsville, this was to be a straight uphill climb. Additionally, Dave Taylor would be coming down to speak at the conference, so Carolyn, seeing that as an opportunity, had come up with a special outreach campaign that she labeled "putting down roots."

Before we got to that, we had a Congressional luncheon at the Von Braun Center to attend first, that being the one that Senator Jeff Sessions annually held in April to let the Huntsville business community know what he was doing to make sure that federal money was headed their way in large quantities. This was actually our first official Huntsville event, and since we were pretty busy getting ready for the suppliers conference, Carolyn, Dave Murrow, and I went, and Carolyn invited Rick Davis of the Chamber of Commerce, Jeff Irons (whom Carolyn knew from her days in Washington, and who seemed to be the unofficial mayor of Huntsville), and a couple of other potential local suppliers to our table. Carolyn managed to get our group picture taken, so we got some mileage out of that as well. But mainly we began to be seen around the community.

Carolyn's putting down roots campaign was clever. She knew from me that Mayor Spencer had invited Dave Taylor back to Huntsville to tour the Huntsville Botanical Garden, so Carolyn wanted to take advantage of that. With Dave coming to town for the April 27 suppliers conference, she thought that would be the perfect time to do something public with the mayor. Her idea was for the mayor and Dave to plant a tree together to signify Ball Aerospace putting down roots in Huntsville. She wanted to use a Colorado tree to plant in Huntsville, but given the hot and humid Huntsville summers, that was a bit of a challenge. An Aspen tree definitely wouldn't survive, so she consulted with the Botanical Garden botanist, Harvey Cotton, who eventually recommended a Rocky Mountain skyrocket juniper. Yes, there actually is such a tree, but it wasn't available in Huntsville. Harvey was able, however, to locate one in Birmingham, and then we were set.

But first, there was the suppliers conference. Other than the administrative assistants, Carolyn was the only nontechnical person assigned to the Ares pursuit team in Huntsville. Hopefully you can imagine what it's like for a nontechnical person to have to work closely with a group of highly technical engineers. The engineers' hearts were all in the right place, but beyond that, it was tough. First, she wanted nice invitations printed for the conference, like you would have for a wedding. Then she engaged a local caterer to provide some nice hors d'oeuvres and beverages. And then there were the flowers, and shipping in pictures of hardware that Ball Aerospace had built to hang on the walls, and spacecraft and launch vehicle models to sit around strategically. And she made everyone dress up in a suit, or at least a coat and tie, or a nice pants suit or dress, as appropriate. Plus, we rehearsed the event as well. We also had a half section of a full-scale Instrument Unit Avionics ring coming by truck from Boulder. The engineers all wanted to crowd our space with this, but Carolyn insisted that we only use a quarter section, which then would nicely serve as a photographic backdrop for the various speakers at the conference. When the day arrived, everything was in pea-pumpkin order (that's a southern term!). And everything went very well. We had about 125 people show up from the local business community representing about seventy-five companies. We knew that it was a hard life for a second-tier supplier in Huntsville, as we had been told that when large companies located in Huntsville, for example Boeing, they simply dictated terms and conditions to the little guys. Our approach was to be more of a partner, and to work together with them as a team. It was a well-received idea, and as a consequence we picked up many potential suppliers that day, such that the conference was a huge success.

After the conference it was time to get the putting down roots campaign into high gear. First, there was a luncheon for Dave Taylor with the mayor and several other community leaders, including some local congressional staffers. It was held at the Ledges country club up on the side of one of the local mountains that overlooked Huntsville. One of the very funny things that happened in that luncheon involved me. I mentioned to the Mayor that we had signed up to cook at the upcoming Rocket City BBQ, a large three-day festival held in early May each year. She asked me if we were pros at this, and I said, "Absolutely not, and that's why we registered with the shade tree [amateur] group." So far, so good. She next asked me if we were cooking wet or dry. Of course, I had no clue, but I seized the opportunity for her to teach me what the difference was, so I guess that worked out okay. In any event, everyone at the luncheon got a good laugh out of that.

After the luncheon we all headed to the Huntsville Botanical Garden for the tree-planting ceremony. Everything was all ready to go, and the *Huntsville Times* was there to take pictures and do an interview with Dave, so we were guaranteed good local publicity. Our tree got planted in the Space Garden, a subsection of the full Botanical Garden, and very befitting of a city known as Rocket City. The planting ceremony went really well, but we did get a little surprised by the Botanical Garden having decided, without discussing this with us, to give Dave a tree from Huntsville to take back to Colorado. It was a dogwood tree sapling with a story. It seems there had been a local 125-year-old dogwood tree that was threatened with being dug up, and in the process, killed, to make way for a new road. But the schoolchildren of Huntsville started gathering up their pennies, nickels, and dimes and eventually raised almost $15,000 to have the tree professionally moved to the Botanical Garden. Such moves of very old, well-established trees are rarely successful, but this one was, and well after the move it was doing just fine and was absolutely beautiful when it bloomed in the spring. And our sapling was an offshoot of this very tree, which made it quite significant. I asked the botanist if it would survive the harsh winters in Colorado, and he said that it was zone five hardy and would do just fine. Our job thus became to get it safely to Colorado, which we eventually did, but not right away. In the meantime, the Botanical Garden kept it and tended it for us, until we could make the necessary arrangements to safely move it. Thus, not too long after the tree planting event, it was standing proudly outside the Ball Corporation cafeteria in Broomfield, Colorado, with a suitable plaque noting that it was a gift to Ball Aerospace from the city of Huntsville. Imagine, a dogwood tree in Colorado.

Based on a recommendation from our consultant, Parker Counts, we had gotten a corporate membership at the Heritage Club, a country club–like arrangement, but without the golf course, since it was right in downtown Huntsville. We used this frequently for company functions, entertaining current and prospective business partners, and whatever else was needed. This was another good place to be seen. And the membership director, Fran Whitlock, was a genuinely nice lady to work with.

So, the evening of the suppliers conference, Carolyn arranged a big thank-you dinner for the entire pursuit team at the Heritage Club, and Dave and I both spoke and thanked the troops. A genuinely good time was had by all. It always helps team morale to do things like this, especially when they have just successfully completed a big push-up.

The next day, a Saturday, the plan was to videotape Dave and me speaking about how important the Ares I IUA job was to Ball Aerospace, and how well positioned we were to do this job, all in front of the quarter section of the full-scale mock-up of the Ares I IUA ring. This was to be included in a ten-minute video that we were preparing to be able to use with Marshall to help explain how we would do this job. After a couple of takes (Dave wasn't feeling the best that day), we got a good shoot in the can, and it was time to take Dave to the airport to head back to Colorado. It had been an exceptionally good trip to Huntsville for him.

Now we were on to the next thing, which was the Rocket City BBQ, the next weekend. One of our engineers from Boulder, who we knew to be good at cooking barbecue, was coming in for an on-site engineering review, and he readily agreed to stay over for the big barbecue event. It was a three-day affair, starting on Friday and continuing through Sunday, with the main judging accomplished Saturday evening. There was a precursor get together Thursday night before the real thing, which Carolyn and I went to, to wave the Ball Aerospace flag, so to speak.

On Friday the event started in earnest, and we had our beef, pork, and chicken in the rented cooker early, so that it would be able to get the requisite ten to twelve hours minimum of slow temperature cooking and so that the flavor of the basting could soak in. That night we tasted it, and we immediately knew that we had to do something different with the next batch of meat. To the rescue came Vickie Terry, one of our excellent publication folks, who suggested that we add Coca-Cola to the basting to help keep our meat moist and give it some additional flavor. That worked like a champ, the proof being that we actually scored reasonably well, especially as a beginner in this event. But, perhaps more importantly, we beat out one of our well-established competitors, Lockheed Martin, in the final judging.

I should note that this wasn't all just cooking for fun, as the event was open to the public, and we were fortunate enough to have a number of our potential teammates stop by, as well as many folks from Marshall, including some of the key Ares I IUA folks. This ended up being a positive event, with a view being created that we really wanted to be part of the community. Thanks again to Carolyn for getting us into this event and arranging some of the logistics to help make it happen.

Next on Carolyn's agenda were the black-tie events that Huntsville seemed to thrive on. I was used to going to exactly one black tie event a year, and that was the National Space Club's Goddard dinner held in Washington, DC, in the spring of each year. But Carolyn warned me that Huntsville was different, so I brought my tux and hung it on the back of the door to my office so it would be ready to go at a moment's notice. We ended up attending about six such events over the spring, early summer, and fall. I will discuss those that were most important to our efforts as we go along, but suffice it to say that black tie events were always just around the corner. And all the senior people in the business community went to them, so it was essential that we be there too. And Carolyn always saw to it that we could go.

And then there were the many Chamber of Commerce events that ranged from breakfasts to luncheons to simply meetings, all of them important to participate in and be seen attending, as everybody who did business in Huntsville was expected to actively participate in and sponsor these events.

Working with Boulder, Carolyn also facilitated the placement of advertisements at the airport so that you saw them as you were going to, or coming from, the airport gates. And a big billboard with our ad on it was placed on the road going into Marshall's Gate 1, which most of the NASA folks used to go to work.

So this was Carolyn's job, and she did it exceptionally well, making sure that we were always where we needed to be, that we were dressed appropriately, and that we sponsored as many of these events as we could, so that we could be recognized as active, contributing members of the Huntsville business community.

There is no clearer evidence of this than in October, when the down select (a process the government conducts to reduce the number of bidders to those most likely to be able to meet the requirements of the RFP) occurred and Marshall went from five bidders to two. The story in the *Huntsville Times* was titled "Two Local Firms, Ares I Finalists ..." One of them, Boeing, had been in Huntsville for thirty years. The other, Ball Aerospace, had been there six months. Thank you, Carolyn.

PART 3
The Proposal

Getting Ready for the RFP

The draft RFP came out on April 17, right on schedule. We had been following this very closely, plus we had already totally dissected the Ares I Upper Stage RFP, which had come out earlier, and which we expected to be a model for the Ares I IUA RFP. And it pretty much was, so there weren't very many surprises.

The main changes were as follows:

— The mission suitability section of our proposal was now limited to one hundred pages (it had been 225 pages for the Upper Stage).

— Marshall was providing fifty seats for the colocation of contractor personnel with the NASA civil service team (this was a good change; however, it didn't hold up, forcing us to scramble for additional office space at the last moment, after the final RFP came out).

— During the execution phase, we were to report through the Ares I Avionics and Software Integrated Product Team (IPT), under Marshall's Charlie Nola.

— There was a new requirement that all our key personnel had to commit to stay in Huntsville for a one-year time period (since

Ball Aerospace employees are at-will employees, which means that they can leave Ball Aerospace whenever they wish, this proved to be a significant added difficulty for us and our teammates).

- The cost volume appeared to be particularly onerous.
- Orals after the submission of the proposal had been added; this is where we would have a chance to present the essence of our proposal to the source evaluation board (SEB).
- They were saying that there would be no best and final offer (BAFO), but that didn't hold up either, as we had to submit a final proposal revision (FPR; equivalent to a BAFO) after going through a formal questions/discussion set of interactions with the SEB.
- The schedule was still holding firm:
  a) Final RFP release, June 6
  b) Past Performance volume due, June 27
  c) Mission Suitability volume due, July 16
  d) Cost Volume and Model contract due, July 30
  e) Oral presentation, August 3
  f) Award announced, December 1

Industry day was held April 30, which was when Marshall explained their RFP to all the prospective competitors. No one from industry ever asks any questions at these public events for fear of tipping off the competition as to what you may be thinking about, that they aren't thinking about, or are looking at differently. However, just by looking around, you get a better sense of exactly who your competition is, or might be, just from recognizing who is there.

On May 2, I had the chance to take a few of our folks over to Marshall to meet the SEB members for an hour, and to be able to present them with an overview of our company and why we thought we could credibly respond to this RFP. This would be the only significant interaction with Marshall that we would have where we did not use a coach to get us ready. Nonetheless, the interaction went well, although we later got feedback that we spent too much time talking to the SEB about our ability to do fixed-price work. We adjusted our approach somewhat after hearing this.

On May 3, an 8 × 10 lighted sign with the Ball Aerospace logo went up on the side of our offices on Sparkman Drive. Now everyone who rode by could see that we were really here.

And on May 4, we submitted our comments against the draft RFP to Marshall. This is where industry tries to influence the government to change something in the draft RFP that they believe may work against them or be difficult to do. Sometimes this works, and sometimes it doesn't. But it's always worth a shot.

So we continued to have a whole lot of work to do, and time was getting short. We had been in a seven day a week mode for some time, and that would only get worse as time went on because the days had to get longer and longer to get everything done.

I became somewhat consumed with meeting with our customers whenever the opportunity presented itself before the black curtain fell upon final RFP release (this is when NASA would stop meeting with us about this procurement). A sampling of such interactions included me meeting with the following:

- Chris Scolese (then the chief engineer of NASA, and someone who had previously worked for me).
- Bryan O'Connor, the NASA Headquarters head of safety and mission assurance, an ex-astronaut, and someone that I knew well.
- Scott Pace, and others, in the NASA Headquarters Program Analysis and Evaluation (PA&E) group.
- Chris Guidi, the NASA Headquarters program executive for Ares.
- Congressional staffers on the Hill, plus, with Dave, a couple of House and Senate members.
- Dave King, the Marshall center director, who was absolutely insistent that whoever got selected needed to operate out of Huntsville, so as to be close to Marshall; he did not want folks wasting their time on planes.
- Steve Cook (a second time), who is the Marshall director of the Exploration Launch Vehicle project office.
- Dan Dumbacher, the new Marshall head of engineering.
- Robert Lightfoot, the new Marshall deputy center director, who replaced Charles Chitwood who I had met with twice before (Robert is a really great guy who became the Marshall center

director in 2009, and moved to NASA Headquarters to become the number-three person at NASA in 2012, eventually moving up to become the acting NASA Administrator, before retiring from NASA in 2018); this was just before the black curtain went down. I did manage to meet with Lightfoot one more time in July, after the RFP release, as a follow-up to a recommendation that Todd May at NASA Headquarters made to Dave Taylor that we see Robert Lightfoot at Marshall about his ideas as to how Ball Aerospace could more quickly get integrated into the Huntsville community; of course, we did not discuss Ares, but he did give us a contact that led us to be able to sponsor a special showing of *In the Shadow of the Moon* at the US Space and Rocket Center in the fall, while our proposal was being evaluated.

- Danny Davis, the Ares avionics lead who works for Charlie Nola; Danny remembered us from the Rocket City BBQ when he stopped by our booth with his son and sampled some of our cooking; this was the last person at Marshall that I saw formally before the black curtain went down when the final RFP was released.
- Tip Malone at NASA Kennedy, who would be the receiver of the Ares I launch vehicle hardware once it started arriving at Kennedy for launch.
- Jim Hattaway, at Kennedy, who is the Kennedy associate director (number-three person).

We also did the following:

- attended a Shana Dale (NASA deputy administrator) Woman of the Year function in Washington, DC
- attended the Huntsville leadership dinner with Colin Powell (another black-tie affair)
- cosponsored the NASA Administrators VIP reception at the June space shuttle launch (this was another of the events that, with support from Boulder and the Washington, DC, office, Carolyn had been able to figure out how to get Ball Aerospace into); we got to talk to lots of key NASA folks while we were at

Kennedy, including Mike Griffin, Rex Geveden, Dave King, Robert Lightfoot, Dan Dumbacher, and many others

— attended the US Space and Rocket Center Space Camp 25th Anniversary dinner (a black-tie event), with William Shatner, of *Star Trek*/Captain Kirk fame as the MC for the event

— attended a special Ball Aerospace sponsored showing of *In the Shadow of the Moon* in Washington, DC

Lying on top of all this was a pink team review, a red team review, reconvening our external senior strategy review group, plus continuing to have monthly "must-win" meetings with Dave Taylor and his executive staff in Colorado. So my day job hadn't exactly gone away either.

During the pre-RFP release time period, my biggest concerns were our proposed cost, finalizing teaming arrangements, working out staffing details, especially including getting key personnel on board, and getting both Ball Aerospace and Ball Corporation executive approval (required for anything being proposed by Ball Aerospace for more than $100 million).

Cost would be a never-ending concern for all of us, since this selection official, Doug Cooke, had a history of always selecting the low bidder, thus significantly raising the ante on keeping our cost down, but in a credible fashion.

I had the Ball Aerospace/Ball Corporation final proposal reviews scheduled so we at least had a plan for accomplishing that piece of the puzzle. Getting the right words on paper for these sorts of events was always a challenge, but it always came together in the end, so at least we now had a schedule to push against.

My teaming and key personnel concerns had to be resolved sooner, however, so these became additional areas of significant emphasis on my part, and they were significant enough efforts that I have devoted a chapter to each coming up next.

Teaming

Finding the right teammates was a critically important aspect of being able to win this job, and it was one that we spent many months diligently working on. As should be obvious from the chapter 7 discussion about our competition, we were the little guy on the block, David against several Goliaths, so to speak. It was also fairly obvious to even the casual observer that we would have a lot of difficulty doing this job by ourselves and would probably not be credible if we were to propose such an arrangement.

So what were we looking for in teammates? Simply put, a company or companies that:

- complemented our capabilities, making our proposal stronger as a consequence;
- was very experienced with NASA and in particular the human space flight program, preferably at Marshall;
- was very experienced with building production avionics and in particular, experienced with building aircraft avionics;
- wasn't so big, or known to be so argumentative, that the reviewers would still have confidence in our ability to manage them; and

- wouldn't push a particular avionics solution/approach, which immediately eliminated Honeywell, since they were actively and publicly promoting that their approach for the crew exploration vehicle/Orion Avionics be used for the Ares I IUA, much to the chagrin of Marshall, who, of course, wanted to develop their own design.

There were other attributes that we were looking for, but these were the biggies.

So who did we talk to? Virtually everyone we thought met at least some of the key qualifications noted above, and who appeared to be interested in this competition, plus a few more.

The following is not an exhaustive list but rather the most important of the companies that we considered and/or talked to about teaming with:

- the United Launch Alliance (ULA) (one problem was that any teaming arrangement with them would have to be approved by both Lockheed Martin and Boeing, which wasn't likely to happen)
- United Space Alliance, but they had no significant hardware experience; they ended up teaming with BAE Systems
- Raytheon, who appeared to be more interested in priming (which turned out to be correct)
- BAE Systems, but they clearly wanted to prime the job
- General Dynamics, who clearly wanted to team with somebody, and we did have some very serious discussions with them, but they eventually decided to team with BAE Systems, mainly because we could never figure out how to divide the work share given that we both brought a lot of the same capabilities to the table
- Teledyne Brown, located in Huntsville, and who had manufacturing capabilities sitting idle, but they were ground-ruled out by Marshall procurement, because of their close contractual connections to Marshall in this general area
- Harris, who appeared to also want to prime, but eventually decided not to do that, and then joined the BAE Systems team
- Orbital, who appeared to have their hands full with doing the CEV/Orion Crew Escape Tower

- Northrop, who had gotten badly burned in the process of losing the CEV/Orion job to Lockheed Martin, and who was now a sub to Lockheed Martin
- Lockheed Martin (but we'd heard they had some sort of exclusive agreement with Honeywell, which must have been correct as they ended up bidding the IUA with Honeywell as a prime and Lockheed Martin as the subcontractor)
- Hamilton Sundstrand (more on them later)
- Pratt & Whitney Rocketdyne (more on them later)
- Boeing, who we had actually first approached at their offices in Huntsville

We had initially approached Boeing in Huntsville in mid-March around the time that the Ares I Upper Stage proposals were due and before the time of the draft RFP release for the Ares I IUA. Amazingly, but certainly indicative of their overall depth and strength, they had not even decided whether they were going to bid the Ares I IUA, and if they did decide to do that, they had not even begun to address whether they would prime it. They said they were currently consumed with bidding the Upper Stage job (which I perfectly understood) and that they needed to see the draft RFP for the IUA before making those decisions. That meant they would not even begin to form a pursuit team until mid-April, and we had been chasing this opportunity since early 2006. But as discussed in chapter 11, at the Brewster Shaw level, them teaming with us, where they would be the subcontractor and we would be the prime, was a total nonstarter.

So that brings us to late April, when we finally struck a deal with both Hamilton-Sundstrand (HS) and Pratt & Whitney Rocketdyne (PWR). Between the two of them they checked all the boxes of what we were looking for in teammates.

PWR was already well entrenched in Huntsville as a Marshall contractor and had almost nothing but human space flight experience. They also brought extensive power subsystem and software experience to the table, both of the human-rated type. And their customer familiarity quotient was off the scale because of their long history with Marshall.

HS, on the other hand, was not located in Huntsville, but they had a great avionics capability and were a major provider of avionics for the aircraft industry—and Steve Cook was always saying that that was what he wanted—a contractor that could migrate the aircraft avionics approaches to the Ares I IUA, so that down-the-road avionics upgrades weren't the huge issue that they had proven to be with the space shuttle. They also brought to the table some very clever, highly proprietary approaches to building avionics more efficiently, and more inexpensively. And they too had extensive human space flight experience, as they had built the astronauts' spacesuits for a very long time.

From our way of thinking, the combination of Ball Aerospace as prime and HS and PWR as teammates was a great arrangement, almost a dream team, if you will, and we were absolutely ecstatic when the dust was all settled on this deal.

Now we were set and ready to move to the next level of this competition with our team fully in place. But next, I still needed to deal with the staffing/key personnel issue.

Key Personnel

Staffing for the proposal was one thing, but staffing for the execution phase was quite another, especially in the key personnel area. Ball Aerospace had pretty much always been mainly located in Colorado, principally Boulder. And having lived there myself for a number of years, Colorado certainly has a lot to offer, from skiing to mountain climbing to hiking to biking to fishing, not to mention more than three hundred days of sunshine. Extracting people from that sort of environment was not easy.

On the other hand, Huntsville was a very attractive setting as well. It didn't have the skiing, and it was definitely hot and humid in the summer, but it did have mountains, biking, hiking, fishing, boating, and perhaps more importantly, a cost of living that was about 20 percent less than in Colorado, mainly because of the lower housing costs. So Huntsville definitely had a lot going for it. But still, it was not easy getting people to agree to pull up their roots and move.

Thus, I made a special, unplanned trip back to Colorado on June 18 to attend Dave's weekly Monday morning staff meeting in person, rather than by telecon. I had gotten time on the agenda to discuss the staffing issue with Dave and his staff. We had a great discussion about what Huntsville had to offer, what we might be able to do to further entice people to move, what our specific needs were, and what we needed first, in terms of staffing. We needed a commitment from these folks, and then a move date.

While the meeting didn't accomplish everything that had been on my wish list, it did pick up the majority of those items, and things began to move fairly well on this front. It would be an area that I would have to watch carefully, and an area where I would occasionally have to prod some folks, but overall, I was satisfied with the outcome of this meeting.

But then there was still the separate and special issue of key personnel, and most importantly, getting an appropriate program manager on board. This one had been an ongoing issue, with its genesis in the time period that we were trying to find a senior executive to lead the effort, where the decision was eventually made that I was that person, since we had been unable to meet the compensation expectations of the person that we had originally been interested in. Ball Aerospace is a wholly owned subsidiary of Ball Corporation, and as such we had tremendous independence from our parent organization. But in the area of employee compensation, we had, of course, to abide by Ball Corporation's policies and rules. Exceptions can and are made, but it is not always easy. Nonetheless, while Ball Corporation is principally a packaging company, they do understand the sometimes-high compensation expectations of aerospace industry senior executives and other key personnel. They work hard to try to meet those expectations if at all possible, particularly where the person in question is vital to the needs of Ball Aerospace. That had been the case with respect to getting an outside senior executive onboard, and it would also be the case in the pursuit of a program manager as well.

As is typical in situations like this, we employed a headhunter to help us, one that we had a lot of experience with, and who usually came through with the goods. So, after arriving in Huntsville, Jeff Osterkamp, VP of our program management organization, and I spent a lot of time interviewing candidates for that position. I met with a lot of really good men and women, but there was one in particular that I thought would be great for the job, a person who was well known and well respected by Marshall. But, in the end, we couldn't match their overall compensation expectations. So, like with the senior executive situation, there was no deal, which forced us to look elsewhere.

Fortunately, our headhunter identified someone else who brought some unique, highly relevant things to the table, including a mix of both aircraft industry and launch vehicle development experience, things that were in keeping with our understanding of what Marshall was looking for. So, in the end, that's who we went with as our program manager.

The RFP Is Released/the Proposal Is Submitted

The Ares I IUA RFP was released on June 6, as planned, thus the procurement remained on schedule. As usual, we searched first for changes in the RFP, this time with respect to the draft RFP. Such changes included the following:

- An increased emphasis on innovations; since Ball Aerospace is seen in the industry as a creative and innovative company, and given what we knew about some of the innovations that Hamilton Sundstrand would bring to the table, this boded well.
- The two Indefinite Delivery, Indefinite Quantity (IDIQ) Contract Line Items (CLINS) remained unchanged, with maximum values of $80 million (CLIN 2) and $340 million (CLIN 4), which was also great, as that was the principal growth area of the contract, once it was awarded.
- The anticipated funding profile was largely unchanged, except that contract year 1 had increased from $1 million to $5 million, which was also good, as that meant we could get off to a quicker start, which we all felt that we needed to be able to do so that we could bond as early as possible with the Marshall team, such that

we became the desired seamless, badgeless team that Marshall had been saying they wanted.

- The requirement to deliver a video had been dropped; this was very disappointing, as we had spent quite a bit of money producing our ten-minute video, and we had built the full-scale mock-up of the Ares I IUA ring mainly for the video; plus the video had turned out well, and we had hoped that Marshall would get to see it. As it turned out, we got to use the video very late in the game when Marshall announced their need for one to use snippets from for the press conference when they announced the contract winner, so at least the video had some use and we didn't have to do another last-minute scramble.

- The oral presentation had been firmed up in that we now had two hours, one for mission suitability and the other for cost.

- The fifty colocated seats at Marshall had gone away (this outcome was the most disappointing of all); this presented a significant problem for us, as now we had to really scramble to find new office space—and find it quickly. While we currently had five thousand square feet on Sparkman Drive (our customer-facing Headquarters) and sixteen thousand square feet on Williams Avenue (where we were writing the proposal), the Williams Avenue space was only available for three months (almost exactly what we needed for the intense proposal writing period). It was located on the second floor of a bank building, and that bank had merged with another bank, and thus was closing this location, after which the building was being renovated. But the Ball Corporation property folks in Broomfield were really good, and they rather quickly identified a very nice twenty-thousand-square-foot location at 4815 Bradford Drive, still in Cummings Research Park, and just around the corner from our Sparkman Drive location. Problem solved, but that was a close one, as we had to demonstrate to Marshall in our proposal that we were ready to start work on day one of the contract, and having a suitable office space was part of that. Ironically, in the end, the fifty seats came back, but by then we had found another use for the Bradford Drive space, so all was still well.

Obviously, we had been working on our proposal for many months already, but some of that was based on a best guess as to exactly what Marshall would want. Now we knew exactly what they wanted, so we could now start wrapping things up. Plus, we needed to allow a couple of weeks to get the various volumes of our proposal printed, and Ball Aerospace and Ball Corporation executive approval was still in front of us too.

Our June must-win meeting with Dave Taylor and his staff focused on the remaining open issues that required executive intervention to resolve them. With that behind us, it was time to bear down on important business decisions, such as what company contributions would be offered up as a way to demonstrate our commitment to the job. The government was always interested in seeing a company put its own money—capital investments, especially—on the line. This is what is known as having skin in the game. So this was another area of significant discussion at the executive level.

And what fee (profit) we would bid the job at was something else that would be sending a strong signal to Marshall about where we were coming from on this job. If we bid high, then it would appear that we were just in it for the money. If we bid low, it might appear that we were trying to "buy-in," plus Marshall would likely feel that they had less leverage over us if we didn't have much fee at stake relative to our performance. So the trick was to find the best middle ground, not always an easy task.

By now it was July 4, which, while it was a holiday for a lot of folks, was just another workday for us. I did, however, encourage folks to break as early as possible and participate in some of the July 4 celebrations (including a minor league baseball game involving the Huntsville Stars) that were occurring around the area. Carolyn and I went up to Ditto Landing on the nearby Tennessee River to partake of some true Americana (including an Elvis impersonator singing "Amazing Grace"; imagine that), plus watch the July 4 fireworks display. It was one of those things that, after having done it, you realized that you wouldn't have wanted to have missed it for anything. But after it was over, it was back to work.

We had a special preproposal review on July 9 with Dave and a few others but not his full staff. We still had a few open issues, and a couple of them, having to do with liability, were pretty sticky ones and would certainly garner Ball Corporation's attention when we went to them for approval of our proposal. Thus, we needed to reach an agreement on how best to deal with them.

The next key step was to have a formal proposal review with Dave Taylor and his staff prior to doing likewise with Dave Hoover (CEO) and Ray Seabrook (CFO) at the Ball Corporation level. The review with Dave Taylor was planned for July 16, with the Dave Hoover review scheduled for the next day.

Unknown to me, our consultant, Parker Counts, had called Dave Taylor in advance of the July 16 review and suggested it might be good for team morale if Dave would come down to Huntsville to conduct the review there. I had no idea that this discussion had taken place, and Dave didn't make up his mind instantly, as he had to move some things around on his calendar to be able to make this happen. So, while this was a really good idea on Parker's part, down in Huntsville, people were packed up and ready to head to the airport to go to Colorado (and in some cases home to see their families, at least for a short while). Some folks had actually already left. Carolyn and I had stopped by the Williams Avenue office to spend some time with the team before leaving.

Carolyn and I were getting ready to head to the airport when Dave called me and sprang this idea on me. So it was scramble time again, as we hit the phones to try to recall folks from the airport if possible. So again, a great idea, but I sure wished I had known about it earlier. Parker later apologized once he realized the confusion it had caused, and he and I had no future miscommunications of this sort.

Dave came down and brought Bill Unger, the Ball Aerospace finance vice president with him. Carolyn and I picked them up at the airport that day in our leased Ford Mustang. Bill Unger is a really tall guy, so getting him and Dave into the back seat proved a challenge—so much so that Bill anointed our Mustang the "clown car," which is how it would be referred to forever.

Everyone else participated in the review via telecom from Dave's conference room in Broomfield. The review went well, and we crossed the necessary t's and dotted the i's, and everyone felt good about where we stood with the proposal. After that, Dave Taylor, Bill Unger, and I reviewed the draft package that I planned to use with Dave Hoover, and we fine-tuned that as well. Dave also spent some quality time with the troops, including having lunch with them over at our Williams Avenue office.

Now, it was really time to go to Colorado! And this time we (a much smaller contingent) did make the trip. I made the presentation to Dave Hoover and Ray Seabrook, and it was well received overall, and Hoover and Seabrook said that they were satisfied with the way that we had dealt with the various risk elements of our proposal, such that we didn't leave them sitting on a limb that was rapidly being sawed off. Dave Hoover is a really great guy who I thoroughly enjoyed working with during my time at Ball Aerospace. He is one of those kinds of folk who fit right into any environment, whether it is sitting down with some of the working troops to have lunch in the cafeteria or meeting with a senator or member of the House of Representatives. He's quiet and unassuming, yet very sharp, and extremely business savvy. All in all, someone that you very much want to have on your side. And he was totally behind what we were doing, so that was great.

After the review, I sent a note to the entire pursuit team telling them how well everything had gone and passed along Dave Hoover's personal thanks to everyone for having worked so hard while, in many cases, being away from their families. They appreciated that very much.

Carolyn and I had to rush back to Huntsville immediately after the Hoover review because I was supposed to represent Ball Aerospace at United Space Alliance's (USA) conference in Huntsville, where we were to be presented with their Large Business Supplier of the Year award on July 19. It is a little-known fact, but one of those things that we promoted to Marshall whenever we got the chance, that Ball Aerospace built the star trackers that are used to navigate the space shuttle. And we've been doing so since the very inception of the space shuttle program. We also built the keel latches using in the space shuttle payload bay, and most importantly, the Propellant Reactant Storage Assemblies (PRSA), which store the liquids that are used to fuel the space shuttle's fuel cells which then provide electrical power for the shuttle, as well as oxygen for the astronauts. So, despite whatever you may have heard or thought, Ball Aerospace is right smack-dab in the middle of the so-called critical path for every space shuttle mission. Thus, we can definitely be depended on to deliver reliable products that work well.

The award that we were to get from USA recognized the excellent work that Ball Aerospace had done over the years in the star tracker area. This was great timing, as the black curtain was down, and most of the many Huntsville evening functions that normally occur had ceased for the summer, so there weren't that many venues where we could be seen. To be able to get an award that both signified us as a large business, and cited our excellent work in the human space flight arena, was outstanding in and of itself, plus the timing was almost perfect.

After the dust settled from this excellent event, it was pretty much just turning the crank to get the cost proposal printed and delivered to Marshall on time. The past performance and mission suitability volumes had already been delivered, and consequently many of the folks that had been directly involved in their preparation had already returned to Colorado, so that we could get them off our pursuit payroll, so to speak.

But the huge cost volume wasn't due until July 30, and that set of folks was still working 24-7 to make it happen. The volumes were all cross-linked, so changing something in one of them meant changing it in several of them, so we tried to keep that sort of thing to a minimum, but some changes simply had to be made. I can remember one night when we had hoped to get out early, but we discovered an error that had to be corrected, and it involved working with folks in Boulder, so this one took some time. At first, we were all in denial and thought that we would still be able to get out early enough to be able to go out somewhere to eat. But then it dawned on several of us that that wasn't going to happen. Connie, one of our admins, had the forethought to ask us about a half hour ahead of our usual "bring it in" food sources all closing on us, if we wanted dinner brought in (yet again), so we were able to get Subway meatball subs for all. Yet another gourmet delight.

Finally, it was July 30. As with the earlier volumes, we used a pickup truck belonging to one of our managers to deliver the many, many copies of the cost proposal that Marshall had requested. Talk about killing trees. And now the die was firmly cast, and our fate was, at least initially, in the hands of the Marshall SEB. But the oral presentation was also looming just around the corner.

The Oral Presentation

Around the same time that we were delivering the final cost volume of our proposal to Marshall, Carolyn and I, following up on a recommendation that I had gotten from Robert Lightfoot, met with Shar Hendrick, manager of government and community relations at Marshall. During this meeting I was able to connect the dots back to a conversation that I'd had with Dave King back in the spring. Dave had told me that he was anxious to have a new voice in Huntsville, such as Ball Aerospace, to help deliver the Exploration message. At the time, I read this as another endorsement of the message that we were hearing over and over from Marshall at all levels, which was that they really wanted new blood to win the Ares I IUA contract. Today, however, I saw it as a way to get Ball Aerospace more out into the community during the period of time that our proposal was being evaluated.

Thus, our conversation with Shar focused on two key points: First, I volunteered to help with the planned October 22 Dave King/ Charles Elachi (the director of JPL) Conference on the Crossroads of Space Transportation and Space Science; this conference was planned to demonstrate how exploration and science could peacefully coexist at

NASA. Since I knew Elachi extremely well, having worked closely with him since the 1970s, I knew I brought something unique to the table; however, this conference never really got off the ground, at least at this time, so this ended up being a dead end.

Second, and this is the surprising one: Shar brought up with us, not the other way around, that he really liked how we had been sponsoring showings of *In the Shadow of the Moon* around the country, and he wondered if we could help Marshall do one here in Huntsville (it had been shown at Marshall, but everyone didn't get to see it, nor had it been shown in the community). We had been trying unsuccessfully to make this happen for months, and now Marshall wanted us to help them do it; of course, I said yes, and we went from there. I was skeptical that this idea would actually take root and grow, since the competition was still ongoing; in fact, it changed character to be more of a Ball Aerospace event, but Marshall stayed involved all the way, and we got to show the movie and have a big reception at the US Space and Rocket Center on September 27, with Dave and a number of his staff present. Overall, it went well, and managing to pull off this whole deal was actually quite amazing. Kudos once again to Carolyn and to Sarah Sloan, Carolyn's boss in Boulder, who made all the showing arrangements with the movie company.

At about the same time, Carolyn and I were invited to attend a reception at the Huntsville Botanical Garden that was being hosted by Teledyne Brown as Rex Geveden took over as president of Teledyne Brown Engineering in Huntsville. Dave Taylor and I had met with Rex back in March when he was still the number-three person at NASA. In that meeting I had figured out that he wasn't that enthused with being in Washington, DC, as opposed to Huntsville, plus we had talked at one of the space shuttle launches with his wife, Gail, who also seemed anxious to get back to Huntsville. Thus, this meant that any leanings toward us that Rex may have had were now lost, as far as the selection process went.

After submitting the cost volume, we had quickly become totally focused on the upcoming oral presentation. The date had changed from August 3 to August 7, which was merely a consequence of actually scheduling the five teams who would be making oral presentations to the Marshall SEB. We were to show up with only ten of our key people, and then we would have an hour to present our mission suitability volume, followed by an hour to present our cost volume. This was to be a one-way exchange. We could talk to the SEB, but they would not be asking questions. Rather, they would be in listening mode. We also could not introduce or discuss anything that wasn't already in our proposal, so this could not be a marketing event. It could only be about the facts of our proposal. We were also told that they would be running a stopwatch on us, and we would be cut off in midsentence if we ran over our individual one-hour time slots. The event was also to be videotaped.

We knew that this was a really big deal, and a unique opportunity to explain our proposal to the SEB, so we had been getting ready for this for some time. Now we had to finalize things and get ready to do the actual presentation. Wanting to put our best foot forward, we engaged a coach. His name was Rick Davis (a different Rick Davis than was helping us at the Huntsville Chamber of Commerce). He was a local person with quite a bit of specific knowledge about doing business with Marshall. He was most helpful in getting each member of the presentation team focused on how best to tell his or her portion of the presentation.

We also established a "murder board" (a group of experienced subject matter experts [SME] who asked us very tough questions so that we would be better prepared to answer them when we were asked them for real) to do a complete dry run of the presentation. This was to be a full-dress rehearsal, since Carolyn insisted that everyone dig their suits and ties out of their closets and wear them for the dress rehearsal. A funny thing happened in this regard. We were afraid that we might have some difficulty using the AV equipment that Marshall was making available to us for this presentation, so we wanted to have our IT guy with us. His name was Jeff, and Carolyn suspected that he might not even have a suit, as he typically wore jeans and a T-shirt, as IT people are prone to do. Upon talking to him, she discovered that all he had was his "funeral" suit, which of course he wore to funerals. So Jeff had his wife overnight it to him, and when the time came, he was very spiffy indeed.

We had managed to summarize our story into about thirty mission suitability charts and thirty cost volume charts, and we had developed a very tight timeline for the presentation. When the day came, we all headed to the meeting place and found it locked up tighter than a drum, with a NASA guard sitting outside the door. We all had to sign in and have our ID checked before we were allowed in. This sort of thing can be intimidating to some folks, but we were well prepared.

The presentation went as close to flawless as you could have imagined, and in his closing comments at the end of the event, the SEB chairperson, Johnny Stephenson, said it was obvious from our presentation that we were deeply committed to this effort and to doing this job in Huntsville. That was great feedback and also meant that he had been listening very closely to what we had said.

There was also a space shuttle launch going on almost exactly on top of the oral presentation, and Carolyn had gone to Florida to be with Dave Taylor, who was attending for the first time since we had started cosponsoring the NASA administrator VIP receptions. Between Carolyn and Parker Counts, Dave got to talk to a lot of the senior NASA folks, including many from Marshall—Dave King, for example—so that was a good event as well. He even got to have his picture taken standing in front of the soon-to-be-launched space shuttle.

Plus, there was a Washington update luncheon at the Von Braun Center with Congressman Bud Cramer on August 8. Ball Aerospace was a gold-level sponsor, so we got suitably recognized for this level of sponsorship/participation during the event. As a gold-level sponsor, we got to attend a preluncheon reception with the congressman, which went well too. Another good high-visibility event made possible by Carolyn (and some money).

And if that wasn't enough, we were in the process of making the move from Williams Avenue to Bradford Drive. The trucks from Ball Corporation were there, and they were moving all the furniture, equipment, and so on. This activity was done very expeditiously and was complete on August 8. And our new roadside Ball Aerospace sign was on its way as well.

With the oral presentation behind us, it was back to business. I had decided to form an executive committee (Ex-Com) composed of myself as the chair, and my counterparts from Pratt & Whitney Rocketdyne and Hamilton Sundstrand. I held the first meeting on August 14 and used that meeting to start getting everybody thinking about preparing for the formal questions that we expected to get from Marshall, and how we were going to need to be ready to reduce our cost if Marshall opened the door for us being able to do that. Both PWR and HS had recently been through a similar process with Marshall, and they told us that the level of effort associated with this next round of activities would likely be roughly equivalent to what we had just gone through when we wrote the proposal. Oh joy!

But, for now, while there was a lot to do, we were basically waiting to hear something from Marshall.

Given that, I recommended to everyone that they try to take a break. Taking my own advice, Carolyn and I slipped away to Annapolis, Maryland, to go out cruising on our sailboat for about ten days. We had a Catalina 34 named Jazzy that we'd had for about ten years, and we kept it docked on Back Creek in Annapolis. So pretty soon we were all provisioned and out on the Chesapeake Bay, trying hard to put Huntsville behind us for a while. Not that I didn't stay in touch with the office, but nothing of any particular consequence was happening.

While anchored in a quiet cove off the Choptank River one night, we even managed to see a total lunar eclipse, although we had to get up about four in the morning to see it. But it was worth it, and it reminded us again why we do what we do.

Boeing Wins the Ares I Upper Stage

While we were out on our boat catching our breath for a bit, NASA announced that they had selected Boeing to build the Ares I Upper Stage. The value announced for the contract was $515 million. At the time, I had two opposite thoughts:

- Good, as there is no way that NASA will award both the Upper Stage and the IUA contract to the same contractor; plus, it gives Boeing a nice piece of work in the Ares program, so there will be no motivation for NASA to give Boeing the Ares I IUA contract, just to make sure that they are part of the program.
- Bad, as this would now give some folks the idea that they can do, contractually, what they have always wanted to do, which is to combine the Upper Stage and the IUA, which Marshall had steadfastly objected to, simply by selecting the same contractor for both jobs; this thought worried me greatly.

After getting back to Huntsville, I had the opportunity to have dinner with some of the executives from Team Ares, the team that had bid against Boeing for the Upper Stage. They didn't see that there was any way Boeing could do the job at the price the contract had been awarded. They were clear that their bid had been much higher than Boeing's. There had also been a story in *Space News* (a weekly newspaper about activities in the space arena) that had said more or less the same thing. And if you read the publicly available source selection statement, signed by Doug Cooke, the same guy who was slated to be the selection official for the Ares I IUA, even he noted that there was significant cost risk for this selection. So what happened here? I didn't know, but one thing that was clear: Doug Cooke continued his pattern of selecting the low bidder.

I went into the office the next day and set up another Ex-Com meeting with my PWR and HS counterparts. I told them I thought we needed to figure out how to substantially reduce our proposed price for CLIN 1 (we had been told by Marshall that the price component of the selection would be solely based on CLIN 1) if we stood any chance of being competitive. They agreed. The folks in Colorado had independently concluded the same thing, so while we had been working in the background on this, we quickly lit a huge fire under this effort to figure out how we could significantly reduce our cost in a credible manner.

Thus, Boeing continued to be front and center in my mind, something that had been the case way too long. All I could see whenever I thought about this was Brewster Shaw's smile, at the time that Dave Taylor and I had asked Boeing to join our team.

In the meantime, an interview that I had done back in the summer with Shelby Spires of the *Huntsville Times* finally came out, plus I did another one with *Space News*, which was expected out soon. So we continued to work hard at being seen.

But Marshall continued to be as quiet as a church mouse, with not a peep.

PART 4

The Down Select

One of Two Local Firms, but the Other One Is Boeing!

Marshall never told us whether there would be a down select process. Having chaired an SEB myself much earlier in my NASA career, I always had felt that, with five proposals, they would do a down select as early as possible, simply to reduce their workload. Writing proposals is a pretty intense activity, but so is evaluating them, and the SEB was being watched like a hawk as far as maintaining their selection schedule was concerned. Thus, it made perfect sense to me that there would be a down select. The only questions in my mind were when and how. Would they just do it based on the proposals as submitted, or would they take everyone through some sort of a question and answer process, and then do it?

They also never said whether there would be a best and final offer (BAFO) process, and the current NASA procurement policies suggested there didn't have to be one. When I was the deputy center director at Goddard, there had been a big push from NASA Headquarters to streamline the selection process by selecting the contract winner based on the proposals as submitted, thus avoiding a lengthy down select process. At Goddard, we had been pretty diligent about implementing this change, but had Marshall? I didn't know. So we were a bit in the dark about what was going to happen next. Several of our folks were absolutely convinced that we would get questions for clarification soon, and certainly before a down select, if there indeed was to be one. But after almost two months since our oral presentation, we hadn't gotten any questions—nor heard anything from Marshall, for that matter.

It was October 4, and Carolyn and I had been in Washington, DC, for some meetings. We decided to pop over to Annapolis and spend the day at the Annapolis (in the water) Sailboat Show. I was on the foredeck of a beautiful Beneteau Oceanis 40 sailboat when I got a call around four that afternoon. The down select had been accomplished, and we were still in the running. Carolyn and I were ecstatic. I immediately called Dave Taylor, who was also in the Washington, DC, area, but got his voicemail, so I left a message. At the time I called Dave, we still didn't know who else had made it through the down select gate. Marshall, of course, told us nothing, so we had to figure it out for ourselves.

We started calling our competition and pretty soon the scenario emerged. BAE systems was out. Honeywell/Lockheed Martin was out. Raytheon was out. Boeing was in. Right then and there, since we had both been worried about Boeing for a long time, Carolyn said, "I hope NASA isn't using us to get to Boeing."

But we had just beaten out three excellent competitors, so we were very happy. All of a sudden it was also clear that we were being seen as highly credible in this competition. Boeing might still be around, but everybody else was gone, so obviously we must have done something right. Dave tried to call me while I was on the phone with Huntsville trying to figure out who was in and who was out, so he left a message of congratulations, which, of course, I immediately passed along to the pursuit team. It was a good day indeed.

On October 11, the *Huntsville Times* ran a story about the down select, and the headline read: "Two Local Firms Ares I Finalists … Ball and Boeing are Contenders for the Instrument Unit." The article also referred to Ball Aerospace as a leading aerospace company. You can just imagine how good that made Carolyn feel, as that had been her sole objective in coming to Huntsville to support the pursuit (getting Ball Aerospace quickly integrated into the Huntsville business community). She, with the support of Boulder, had clearly accomplished that in spades.

Questions and Weaknesses

The next day we got a ten-page letter from Marshall that posed questions for clarification and described our weaknesses. We had until October 22 to respond, so we were off and running. We had anticipated as many weaknesses and questions as we could think of, and many we had thought of in advance, so that gave us a nice head start.

I immediately held another Ex-Com with HS and PWR to go over what we had gotten from Marshall and to talk about how we should respond. We also set up a red team process to review our proposed answers. Of course, the discussion turned to reducing our price. We felt that the nature of some of the questions and weaknesses gave us a legitimate license to come back with cost reductions, so we really focused on that. We had also heard a well-placed rumor that our cost had been credible, but Boeing's was more attractive (lower). Who knows whether that was true? But it fit the way we were already thinking, so it just served to reinforce the need to do what we were doing.

Given where we now were, I developed a modified win strategy and started circulating it for comment. It was generally well received, but one aspect of it led to considerable discussion. I said in this strategy that we should reduce our cost, but only as long as our cost credibility could be maintained. I am a person of integrity, and I could see this no other way.

But that's not to say there weren't some folks who thought we should just lower our price and figure things out later. In the end, Dave Taylor, Bill Unger, and I had a most interesting discussion in the Huntsville office conference room about the degree to which the price reductions that we had developed to date were fully and credibly supportable. This led to a more detailed discussion of the rationale for each major element of these price reductions. After this lengthy discussion, I knew Dave and I were on the same page. Given the high integrity of Ball Corporation, this was no surprise, but I was glad to see it verified in practice, as well as in words. So the marching orders to the troops continued to be to reduce cost, but only through clever and innovative ways that were credible and defendable. I then pushed for more company investments and further adjustments to our proposed fee.

And then something happened that pulled the rug out from under me. I thought we were on a good price reduction track, and I was certain that what we were doing was credible and defendable. But on a Sunday afternoon, October 21, I was doing a "wall walk" in our war room on Bradford Drive. We had every weakness and question posted on the wall. We also had every response, in whatever form it existed at the time, posted on the wall. Plus, the cost uppers (there were some) and cost downers were posted with their response. I had been given a net bottom line for the various uppers and downers that I was very satisfied with.

So, I was walking around the room, reading each response yet again, and in some cases challenging the authors to say or do something differently than what I saw on the wall. Then, for some unknown reason, I started to walk the wall mentally summing up the cost uppers and downers as I went. I was sixty-one at the time, and I had grown up in an era where you actually had to know how to add, subtract, multiply, and divide. I had also gone through engineering school using a slide rule, so I had also been taught how to estimate an answer, so that when you got an answer that was more exact on your slide rule, you instantly knew whether it made any sense. Today's kids, in my opinion, have no clue about any of this. This is not their fault, but rather that of our hurry-up society, so that all they know is to use a calculator, or plug numbers into an Excel spreadsheet. Sometimes the result is garage in, garbage out, and they don't know it.

So I completed my wall walk, and I had come up with a number that was about $10 million higher than what I had been told. I went back and did it again and got the same result. Then I asked Dave Murrow to do the same thing, but with a calculator. He came up with a number that was $12 million more than where we thought we were. We did this some more using different people and techniques, but there was no avoiding the fact that we were not where we wanted to be, yet we had to submit our responses the next day. We worked through the night, but there were no more rabbits that we could instantly pull out of our hats, so at the end of the day, we submitted it the way that it really was, and I was again worried that our price was too high.

Once again, there was a space shuttle launch that was laying on top of this exercise, so the only person from Huntsville that we could spare to go to the launch was Parker Counts. The launch went well, and Parker got to talk to Brewster Shaw. Parker told us Brewster had said he was "trying to work with NASA." We had no idea what that really meant, but it reinforced the fact that an incumbent had unique access to their customer. It is absolutely essential that a current contractor be able to talk with NASA, regardless of the fact that there may be a competition in another area going on. They don't talk about the competition that's in progress, but they do talk, and this always gives incumbents the inside track, because they are able to work on their relationship with the customer, and they get to sit in meetings and walk the halls of NASA, where they hear things. If you're on the outside looking in, like Ball Aerospace was, you don't get to do this, so you have to look for other ways to accomplish the same thing. Attending functions around Huntsville helped, sponsoring the special showing of *In the Shadow of the Moon* helped, running ads in *Space News* and the local paper, and doing interviews, all helped, but a nonincumbent has to work much harder at this.

In this vein, the annual National Space Club Von Braun dinner (another black-tie affair; equivalent to the Goddard dinner held in Washington, DC, every spring), was on the evening of October 22, the day that we were to submit the responses to Marshall's questions and weaknesses. Ball Aerospace attended the Von Braun dinner in full force with Dave and a lot of his executive staff present and accounted for. So again, as a nonincumbent, you just do what you have to do.

Discussions

The next step was entering into discussions with the Marshall SEB starting on October 29. As we understood the process, they were doing discussions with us on Monday, Wednesday, and Friday, and with Boeing on Tuesday, Thursday, and Saturday. Discussions were intended to give us a chance to propose corrections to the weaknesses, answers to the questions, and give the SEB members a chance to ask us questions in real-time, and generally interact with us on what we were saying. The idea was that, at the end of the process, all the questions should have been satisfactorily answered, and the weaknesses should have all been corrected, so that both bidders had fully compliant proposals. Given this, the only differences would then be the respective strengths of each proposal—and the price. At least this was the theory. We were to continue doing this until it was done (there was no scheduled end point). But everybody was well motivated to move this process to completion as quickly as possible, since the last thing you wanted to do was to tie the SEB up for more time than they wanted to be tied up.

We set up the war room in our building on Bradford Drive as a command post for interactions with something called the green room at Marshall. This was a room that we and Boeing each staffed on our scheduled day at Marshall so that our discussions team would have a place to retreat to and be able to instantly connect with the rest of the team back at Bradford Drive via phone and fax and an internet connection that Marshall provided. Thus, the discussions team was able to ask a question of the rest of the team in the war room on Bradford Drive and get a written response back suitable to give to Marshall. Plus, the discussions team would be able to have the appropriate conversation with the rest of the team in the war room, such that the team could then intelligently present it to Marshall and be able to answer questions about the response. Each day we also brought a pickup truckload of important documents into the green room (and hauled them out at night), such as the RFP, our proposal, cost backup material, and so forth. Interestingly, Marshall told us that they "swept" the room electronically every night, so that neither we nor Boeing could leave any listening devices in the room. Knowing Ball Aerospace as I do, I knew that would have never been an issue, but I was glad to hear that they did this nonetheless. In any event, we knew this whole process of discussion was going to be pretty intense.

We again turned to a coach, Dr. Rita Kirk, to help us prepare. She was a Professor at Southern Methodist University (SMU). At the time, she was heavily engaged with CNN, helping them with analysis of the presidential campaign, but she was able to carve out a few days to work with us, which we greatly appreciated. Plus, our teammates, who had gone through this process before, staged a mock "day in the life of an SEB set of discussions," complete with dirty (some very dirty) questions to help us prepare better for the unknown. On the Saturday before we started discussions with Marshall, we did a complete full-blown dress rehearsal of day one. And then we did it again on Sunday.

Rita was great. She knew nothing about our business, but she knew everything about how best to present oneself. We could not have had a better coach. Kudos to her, for sure.

She felt it would be important for each of us on the discussion team to try to connect on a personal level with the members of the SEB. Thus, she sat with each of us individually to figure out a little vignette that we could tell during the introduction portion of the presentation, something that would demonstrate our personal interest and commitment to what we were proposing to do in response to this RFP. Watching the reaction on the faces of the SEB members the day we actually did this made me realize that she was absolutely correct. Once again, nonincumbents have to work harder.

Day one of discussions, Monday, October 29, wrapped up late in the day. Overall, it had gone well, and the first-day jitters were gone. Everybody was fully in gear. We had done a two-hour oral presentation to the questions and weaknesses in the morning, and that had gone well. We managed to close almost half the weaknesses, and about two-thirds of the questions had been satisfactorily addressed. But, of course, those were all the "low hanging fruit," so the real work remained.

Day two of discussions was October 31. Of course, the SEB had now met with Boeing for the first time as well, so we were looking for signs as to how that had gone, as possibly evidenced by how they dealt with us today. However, the environment remained friendly and upbeat. We made more progress on closing questions and weaknesses, but clearly things had slowed down. We were also given an amendment to the RFP that established December 12 as the authority to proceed (ATP) date for the contract. Marshall clearly must have felt that they could push Ball Aerospace and Boeing through the discussions process in fairly short order, or they wouldn't have agreed to such a near term date. Or perhaps they were given the date?

Friday, November 2, was day three of discussions, and it was a big day for us. The entire day focused on something called Streamlined Avionics Framework (SAF). This was a major innovation that we were bringing to the table, and made possible by our having teamed with Hamilton Sundstrand (HS). SAF was adapted from a highly proprietary process that HS uses to build aircraft avionics more efficiently and at lower cost. On the surface of it, it seemed to be just what Marshall wanted, given all that Steve Cook had said about his desire to see innovative aircraft avionics techniques migrate into the Ares I IUA world. However, it was new to Marshall and thus would require a lot of discussion with them to get them to understand it and accept it. So our task was to get them fully on board, so that the cost reductions that we had recently taken could be legitimized. At the end of day three, we thought we were in a situation where the SEB got SAF, and clearly understood the benefits to the IUA of using this approach. However, they still weren't convinced that the proposed cost reductions associated with using SAF could be fully realized. But the discussions environment remained friendly, which was a good sign. So we had more work to do, and the weekend would have to be focused on that.

On Sunday, November 4, I called Dave at his home in Colorado with an idea that had come out of a telephone discussion that I'd had on Saturday with one of our consultants, Roger Roberts. I told Dave I was still worried about the infamous $12 million upper hit that we had taken just before we submitted our responses to the questions and weaknesses. I wanted to talk to him about the subject of company contributions and fee yet again. And if we could come to some accommodation on that, would he and Bill Unger be willing to come to Huntsville to meet with the SEB to personally present those reductions in cost to the SEB? As far as coming to Huntsville was concerned, Dave was ready to do that, as he always was. But the other things would take some time. I told Dave that I had the time if he did. So, after spending considerable time with Dave discussing the various options, we came to an accommodation that I was ultimately pretty satisfied with.

Included in this was a totally new idea of turning part of the Bradford Drive building into a conference center. This possibility had risen after Marshall had changed their minds yet again and had offered the contract winner fifty seats on-site at Marshall. Thus, we had leased more space than we actually needed. The idea was to use this extra space by reconfiguring the building to provide convenient access for local government representatives to their subcontractor community, via an off-site conference facility. I haven't mentioned this before, but getting into Marshall was a challenge, given that the army, as part of Redstone Arsenal, ran all the gates. The access rules were as if one was going in to do business with the army, not NASA. Of course, as was the case most everywhere, everything had tightened up considerably after 9/11. We thought providing a place to meet off center would be attractive. Plus, we would propose to Marshall the idea of using this facility to train them and our suppliers in the finer points of how best to apply SAF techniques. So this was to be part of Dave and Bill's pitch to the SEB when they came here on Wednesday.

Monday, November 5, was day four of discussions, and was another very busy day. We made a lot of progress closing things out so that less than ten open questions and weaknesses remained at the end of the day. However, one of these was associated with SAF, and it was still hanging out there wide open, so we were focused on preparing a new presentation to use Wednesday morning to try yet again to address all of Marshall's concerns about SAF. We also made arrangements with the SEB for Dave and Bill to join us at Wednesday's discussions session with the SEB.

On November 7, day five of the discussions, we made our special SAF pitch, but the SEB wanted to think about it before engaging us further, so at that time we didn't know where we stood on SAF with the SEB.

Dave and Bill came down to Huntsville and joined us with the SEB on November 7. Of course, we got to do the whole "clown car" thing at the airport one more time, so that was good for a laugh, which we probably all needed. I had also talked to Dave about showing his personal commitment to the proposed effort by telling a story to the SEB about how he got interested in space when he was a kid, and how that interest had continued to grow over the years. I introduced Dave and Bill, and then, after telling his personal story, Dave talked about how important this program was to Ball Aerospace. After that, Bill talked more specifically about the company investments that we had just decided to make, including the new conference center, and of course, our fee reductions. I felt that it went over well, but I couldn't tell whether it helped a lot, or just a little. But we were clearly doing everything that we could think of to position ourselves for a win, and that was the whole point.

Unfortunately, after Dave and Bill left the SEB to head back to Colorado, we got some bad news on SAF. The SEB said that we had been unable to convince them entirely that SAF would result in the full cost savings we'd said it would. Therefore, that area remained a weakness. They were clear that the only way to eliminate this weakness was to raise our price. I was loath to do this, but we didn't need to have a significant unresolved weakness hanging out there at the time that the SEB presented their findings to the selection official, Doug Cooke. I didn't sleep well that night at all.

The next day I talked to Bill Unger about this, who then talked to Dave, who was tied up with a customer meeting on another project, and we ultimately decided to raise our price to eliminate the weakness. I hated doing this, but part of my logic was that if we didn't, we would still have a significant weakness, and then Marshall would probably add whatever they thought was needed money-wise to their most probable cost estimate anyway. So we could either up our bid price, or they would up their most probable estimate of our cost. Damned if we did, and damned if we didn't. But I felt that it was imperative to at least eliminate the weakness.

Day six of discussions, Friday, November 9, was the last full day of discussions, and this was when we presented our revised cost. Marshall told us that this last weakness having to do with SAF was now closed, as were all other weaknesses and questions. That was good, but our price had gone up, which I knew wasn't necessarily a good thing, but at least we had no weaknesses for Doug Cooke to grab on to.

That night we all retired to Humphrey's, a bar and grille in downtown Huntsville to celebrate the end of discussions. Tomorrow, Saturday, it was back to work on our final proposal revision (FPR). And the clock kept ticking.

The Final Proposal Revision

Late on Wednesday, November 7, after Dave and Bill's visit to Huntsville to meet with the SEB, we were told that our final proposal revision (FPR) would be due November 21. I thought that was a nice touch from Marshall, since that meant that we would be done the day before Thanksgiving, and they would be the ones working over Thanksgiving, and not us. But I think the reality was that they were being pressured by their management about their schedule. In any event, we had two weeks to make it happen. And they were now saying that selection would be on December 12, so we were getting much closer to the end.

The trick with the FPR was to make sure that it incorporated everything, either directly, or by reference, that we had done during discussions. But we had a great pursuit team, and I wasn't worried at all about our ability to do that part of the job correctly. I felt that we remained reasonably well positioned, but we still needed to worry about what Boeing was doing.

And, of course, since our price had changed, we had added company contributions, and we had reduced our fee. I had to get Ball Aerospace and Ball Corporation approval before we could submit the FPR. This review was held in Broomfield, Colorado, on November 15 and included Dave Hoover, Ray Seabrook, Dave Taylor, and me. I made the pitch, and it went very well. Dave Hoover, as expected, understood and agreed with our having raised the price to eliminate the final weakness, and that was that.

In this meeting, Ray Seabrook said something most interesting: "Win or lose, this is the kind of work that Ball Aerospace needs to be going after, so that it can continue to grow."

I felt that he understood full well that this wasn't a sure thing by any stretch of the imagination, and I appreciated that.

The FPR went in on time as expected, and there were no significant issues with getting it out the door. Shortly after it went in, I ran into a non-NASA acquaintance who told me he'd heard the SEB's schedule going forward was as follows:

- Dec 5: review their findings with Steve Beale, Marshall's head of procurement
- Dec 6: likewise, with Dave King and his staff
- Dec 10: likewise, with NASA Headquarters procurement and legal
- Dec 11: a full review with Doug Cooke, the selection official, and others
- Dec 12: a press conference to announce the winner

In the middle of all this were two more opportunities to interact positively with our customer: Ball Aerospace's Annual Holiday Customer Party in Washington, DC, on December 4 at the International Spy Museum, and the upcoming space shuttle launch on December 6.

The holiday party was somewhat uneventful, as we hadn't expected any of the senior leaders involved in the selection process to show up, and they didn't. But we did pick up some rumors to the effect that the Science Mission Directorate (SMD) at NASA Headquarters, and with whom Ball Aerospace had a number of contracts, had been asked to provide some background information on those contracts. We also heard that there was a draft press release in review at NASA Headquarters that had to do with the selection.

As far as the launch goes, I really think Mike Griffin jinxed it by saying, at one of the receptions, that things were going really well and that he expected to get this mission in the bag before the end of the year as was needed to keep that part of the program on track, schedule-wise. Of course, not too long after he said that, there was a problem with a sensor on the external tank and the launch was scrubbed. We did, however, have good interactions with a number of the key folks, including Mike Griffin, Dave King, and Mike Coats (director of NASA's Johnson Space Center).

Brewster Shaw and I had one more conversation before selection. We wished each other well.

23

Getting Ready for Day One

Being ready to hit the ground running when we won the contract had been an unending theme of mine for a very long time. I am certain that my colleague vice presidents in Boulder were tired of hearing me talk about this. But it was important that we not stumble coming out of the gate, so I continued to focus on it. In fact, with the FPR submission behind us, it became my principal focus. I had scheduled some months ago an authority to proceed (ATP) readiness review. I felt that being ready was so important that we ought to treat the ATP event just like we would if it were a launch event. This review was scheduled for November 27.

We went over everything we could think of that bore on our ability to get a quick start on this job. That included getting our facilities and other infrastructure ready, bringing our company investments online, getting SAF, and SAF training, rolling, preparing our teammates as well, and the biggie, which had always been a biggie, staffing for not only day one but day thirty and day ninety. Suffice it to say we were in pretty good shape on everything but staffing, where we still had some significant gaps. These gaps were fixable in the short haul, but there was a cost with doing things that way that I didn't want to pay. So it was back to the drawing board, and I decided that we needed a delta ATP readiness review, which got scheduled for December 10, just in front of the December 12 selection date. This time we were in great shape for day one, and while we still had a few shortfalls for day thirty and ninety, overall, we were in pretty good shape. I was confident that the day thirty and ninety residual shortfalls could be easily taken care of once we won and were a real deal. So now it was getting close to time to head to Washington, DC, to find out who would win.

PART 5

The Decision

The Phone Call

So, after all this buildup, we were back to where chapter one had ended—getting ready for the phone call that would tell us whether we had won the competition.

I'm a great planner. Just ask anyone who has ever worked with me, and they will certainly tell you that. So of course, I had written out some of my thoughts that I wanted to convey to Doug Cooke if we won. And I had done likewise for the case where we lost.

When the call didn't come immediately at (or very shortly after) two o'clock, I began to get worried. I knew from my own days as a selection official that NASA always called the winner first. I verbalized my thoughts about this to those in the room, and by and large the reaction was simply to try to assure me that they were probably running behind. I didn't think so, as I also knew these calls are carefully orchestrated.

The call didn't come until 2:20. It had been a long twenty minutes. I literally jumped when the phone rang. I looked at the face of my BlackBerry, and since I had Doug's phone number stored as a backup, it said Doug Cooke was the caller. So this was "the phone call."

I answered it. Everyone in the room was watching me intently to try to discern what I was being told from my reaction. Doug was gracious but also quick in telling me that while we had a great proposal, he had selected Boeing. I continued to listen, but I gave a "thumbs down" to everyone in the room. This was certainly not the outcome I wanted, but the die was now cast. I immediately pulled my notes for a loss out of the stack of papers in front of me and put them on top. When Doug stopped talking, which wasn't that far into the conversation, it was my turn.

I reminded Doug that I used to be a selection official at NASA, and that I understood how difficult these decisions were sometimes, and that I fully understood that there could only be one winner, regardless of how good the competing proposals may be. I told him that we had given this opportunity our absolute best shot, and thus it was very disappointing that we had lost, but that Ball Aerospace still had a lot of exciting work that they were doing with NASA, and that we would continue to support NASA in its future endeavors, to the best of our ability. I told him that we were totally committed to helping NASA be successful in its science and exploration programs, and that, in fact, we saw some new science opportunities potentially opening up with Marshall simply as a consequence of our recent, extended presence in Huntsville. I also complimented Marshall on having been able to maintain its procurement schedule, as that had helped hold our pursuit costs down. I then thanked him for his careful consideration of our proposal, and that was that.

And that was indeed that. We had lost, so it was over. Dave told me he could never have had the positive, upbeat conversation I'd had with Doug after being told that we'd lost, so he was glad I'd been the one doing the talking. I reminded Dave that there was nothing to be gained by poking NASA in the eye. Dave then talked about what a great job everyone had done chasing this opportunity, and that he was very proud that we had gotten this far in a new marketplace where there was a most formidable incumbent that had to be unseated in order to win.

We decided that we would hang around until the four o'clock press conference, just to see what they said. I knew from experience that little, if anything, would be said about the actual rationale for the selection, but I wanted to see the press conference anyway.

While we were waiting, I put together a thank you note to the pursuit team on my BlackBerry and sent that out to the entire team, our partners Hamilton Sundstrand and Pratt & Whitney Rocketdyne, and our consultants. In addition to reminding them of all we had accomplished during this pursuit, I told them in no uncertain terms that I thought that we had done everything that we possibly could to win, that we had left no stone unturned, that even Doug had said we'd had a great proposal, so there was absolutely nothing to be ashamed of. I sent this note out a little after three and then had something to eat while we waited for the press conference.

The press conference revealed nothing about why Boeing had won and we had lost. And it was really hard to watch Brewster Shaw thanking NASA for having selected Boeing for this job. A press release came out almost immediately thereafter, and that's when we got to see what Boeing's cost had been for the part of the job that the selection was based on. I realized instantly how close we had been, so at least we hadn't gotten blown away on our price. The total announced contract value was just under $800 million, and that was a lot of business to say goodbye to.

We immediately requested to see the source selection document (always a publicly available document that describes the rationale for the selection, which is signed by the source selection official). We also requested a debriefing, and it quickly got scheduled for next Wednesday, December 19. I subsequently requested that it be earlier, and it got moved to December 18.

It was now time to head to the hotel, where our plan was to get together for dinner later that evening.

That Evening

That evening, Dave Taylor, Dave Murrow, Carol Lane, Roz Brown, Carolyn, and I gathered for dinner at an Irish pub called the Dubliner, which was just around the corner from our downtown Washington, DC, hotel. It was to be a mostly somber dinner, although I did manage to take a picture of everyone gathered around and eating "humble pie" (really shepherd's pie). More gallows humor, I guess. And none of us were in a drinking mood either, but we needed to eat, so that's what we did.

But something interesting happened in the Dubliner that night. What do you think that the odds are that we would have ended up in the exact same restaurant the evening of the selection announcement as the selection official, Doug Cooke? I would think the odds would have been about a million to one, at least. But we did just that, if you can believe that. When Carolyn and I saw him sitting at a nearby table as we were heading in, I could hardly believe my eyes. So of course Carolyn and I couldn't resist stopping by and saying hello.

When we got over there, we realized this was really a NASA Headquarters/Johnson Space Center group of folks who had gathered for dinner after a very tough day at the office. Doug was as surprised to see us as we'd been to see him. I thanked him again for his careful consideration of our proposal. He was not amused. I also asked him when the source selection statement would be available, and he said it should be ready now, as they'd had to rework it today, and he'd had to resign it. It amazed me that he would have told me this. Why did it have to be reworked? What had been changed? In my forty years with NASA, I'd never heard of a source selection statement having to be reworked and resigned.

In any event, we parted amicably, and Carolyn and I went back to our table. The two Dave's, plus Carol Lane, were equally surprised by what we had found out.

There was another strange thing about this gathering. Nobody from Marshall was there. They were certainly in town in force for the selection announcement, so why wasn't at least one of them with Doug and his Johnson buds?

The light began to dawn on me a bit about this. Perhaps Marshall still remained outside the centroid of human space flight power, despite Dave King's best efforts to break into the inner circle. But who knows? Maybe it was all just a coincidence. At least everything except the fact that Doug himself had said that the source selection statement had been reworked. Again … why was this done?

We never found out.

PART 6

The Aftermath

The Source Selection Statement

A source selection statement is a legal document that describes a source selection official's (SSO) rationale for his or her selection. It is prepared by lawyers, based on discussions with the SSO, and then, once the SSO agrees to its content, signed by the SSO. Under the Freedom of information Act (FOIA), it is a publicly available document and is frequently obtained by the news media so they can better write their stories about whatever selection is of interest to them.

I have been a source selection official at NASA, so I know that job very well. And I know all about the process of preparing a source selection statement. As you may not be aware, an SSO has something legally referred to as "broad discretionary power" when making his or her selection. In simple terms, the SSO can select anybody he or she wants, assuming the proposer has gotten through the "competitive range" gate (i.e., has a good enough proposal to be considered), and as long as he or she can explain the decision. To be fair, sometimes it is not easy deciding what to do, particularly when you are fortunate enough to have multiple good proposals, where the discriminators among them are small. Also, an SSO always has in the back of his or her mind the possibility of a protest, something to be avoided if at all possible. Thus, the SSOs are well motivated to make a selection that is logical and defendable. But the truth of the matter is that SSOs really can do whatever they want, as long as they can rationally explain why they did what they did. Thus, it becomes the lawyer's job to carefully write down the SSO's explanation of his or her decision such that it is logical, defendable, and legally correct. So it is not uncommon for a source selection statement to be written and rewritten multiple times, until it is right in the eyes of the SSO. But I have never heard of one being rewritten after it was signed, which is exactly what Doug Cooke said had been done for the Ares I IUA. My question here would be, if it was, at some point, good enough in the eyes of the SSO to sign it, why did it need to be rewritten after the fact? Perhaps there's a logical explanation, but I don't know what it is.

Another aspect of the SSO's job is that, despite him or her having broad discretionary power, an SSO doesn't exercise it in a vacuum. When the SEB presents their findings to the SSO, the SSO typically has several key staff and advisers present for this briefing, and then the SSO consults with them prior to making a decision. Furthermore, for a major procurement, it would be a career-limiting move to not discuss their planned decision with their boss, including, in many cases, the NASA administrator. This is part of the reason that we tried so hard to be seen by the people in Doug Cooke's management chain, since we knew full well that Doug would probably consult with at least some of them prior to announcing his decision.

So this is a very carefully orchestrated process, one that, legally, needs to be absolutely airtight if and when a protest of the decision is filed, as something that is said, or not said, in the source selection statement, is almost always used as the basis for a protest, if a protest is filed.

As mentioned, we had requested a copy of the source selection statement for the Ares I IUA immediately after the public selection announcement. Plus, Doug had told me the evening of the selection that it was available. But it wasn't.

The request had been made, as it should have been, from the Ball Aerospace contracting officer to the Marshall contracting officer. This was done multiple times, but the source selection statement still wasn't forthcoming. The next morning, December 13, I was really beginning to get agitated about this, as it was unheard of that the source selection statement wouldn't be immediately made available to the losing contractors. I leaned pretty heavily on our contracting officer about this, and he tried again, this time being told that it was coming. When we still didn't have it by one that afternoon, I decided to break ranks and go directly to the Marshall head of procurement, Steve Beale. Steve had told me that if I ever had a problem to come directly to him, so I did.

I called Steve and left a message. I followed up with an email, and I copied Robert Lightfoot, the Marshall deputy center director. I mentioned in this note that I had seen and talked to Doug Cooke the evening before, and that he had told me the source selection statement was available, as it had been reworked, and he had resigned it that same day (December 12). This got their attention. I immediately got a call back from Steve. He apologized profusely, saying it was delayed because his team was en route to Marshall from Washington, DC, as if to say that they couldn't have sent it to us from Washington, DC, because obviously NASA Headquarters did not have any capability to send faxes or emails with attachments! And, of course, there was no one available at Marshall to send it to us! Right.

We finally got it some twenty-four hours after the selection announcement, which was, in my opinion, a totally unacceptable delay. Had they been reworking it yet again? I never would have had that thought if I already didn't know that it gotten reworked at least once after being signed. Again, we were never able to find anything out about what really went on behind the scenes concerning the source selection statement.

With the source selection statement finally in hand, it was time to dissect it to see why Boeing got selected and we didn't. The bottom line appeared to be that we (eventually) lost on performance relative to some past difficulties with managing cost. Highlights of the source selection statement were:

Mission Suitability:

Both Ball Aerospace and Boeing were rated "excellent," with Boeing having a 3 percent higher numerical score. Ball Aerospace had eight significant strengths and fifteen strengths. Boeing had ten significant strengths and fourteen strengths. As expected, neither team had any weaknesses.

Past Performance:

Both Ball Aerospace and Boeing were rated "very good" overall.

Cost:

Boeing's most probable cost for CLINs 1 and 3 was 3 percent lower than Ball Aerospace's, with both Ball Aerospace and Boeing getting a medium confidence factor in terms of being able to achieve the most probable cost (i.e., the government's estimate of what it would cost to do the job the way each Company proposed to do it)

The cost confidence factor for IDIQ work (CLINs 2 and 4) was "high" for both Ball Aerospace and Boeing.

The cost confidence factor for CLIN 5 production work for both Ball Aerospace and Boeing was "low."

Decision:

For Ball Aerospace, it was noted that our SAF approach was our most important strength. At the same time, it was noted that it had never been done in the space sector, so it was felt that there was some additional cost risk.

Here's Doug Cooke's operative phrase, quoted verbatim:

"I applied the evaluation criteria in the RFP in making my final determination, recognizing *all the factors were essentially equal* [emphasis added by me]. Moreover, I was aware that having two first-rate proposals required me to use finer discriminators than normally used to support selections."

Then there's a whole paragraph on SAF, which included the following telling argument:

"Under past performance, I noticed that Ball received a significant weakness in managing cost and schedule on some of its programs. This gave me an additional basis for concern that Ball could encounter problems implementing SAF, and, therefore, obtaining the cost reductions that Ball projected for the IUA."

The source selection statement was ten pages long, but having read it many, many times, I believe the above describes the essence of the basis for Doug's decision. Thus, he was indeed faced with a dilemma. In his own words, he said he had two essentially equal proposals, so he had to look for finer scale discriminators. For Ball Aerospace, this meant the nuts and bolts of our past performance. I now understood the context for what I had heard at the December 4 Ball Aerospace holiday party, which was that the Science Mission Directorate had been asked to provide additional details concerning our contracts. I will be the first to admit that Ball Aerospace has had some cost management issues, but what contractor hasn't? In fact, during one of the many interactions that we had with Mike Griffin during this pursuit, Mike Griffin, then the NASA Administrator, said, about Ball Aerospace specifically, "They are no worse, nor any better, than any other contractor." Presumably the context for that statement included Boeing, which causes me to ask, what about Boeing's past performance?

Mainly based on my days at NASA, and most particularly having to do with their performance on the GOES-N, O, P, Q contract while I was the Goddard deputy center director, I was personally aware of Boeing's occasionally checkered history on both technical and cost-management performance. So this got me to wondering that, if the NASA Headquarters Exploration Systems Mission Directorate (ESMD) had seen fit to reach out to the NASA Headquarters Science Mission Directorate (SMD) for additional information on Ball Aerospace's performance on their contracts, had they done likewise with Boeing? That is, by reaching out to another organization, possibly outside NASA, since most of Boeing's work for NASA is with ESMD, so ESMD would certainly have direct access to all that information? In particular, was Boeing's performance on a program called the Future Imagery Architecture (FIA) examined by the SEB or, since he had stated that, because he had two essentially equal proposals, he had to look for finer discriminators, did Doug Cooke ask for such information himself, as apparently was done regarding Ball Aerospace's past performance?

The FIA was the National Reconnaissance Office's (NRO) initiative to define, acquire, and operate the next generation imagery satellite architecture (i.e., spy satellites). FIA was supposed to be a constellation of satellites that would gather clearer and more frequent images—even at night and when there is cloud cover—of enemy military activity than current satellites can. Boeing was awarded the contract for FIA in 1999. Boeing's implementation of FIA ran into problems right out of the gate, and in late 2005 (recently enough such that FIA should have been considered in NASA's past performance evaluation), NRO cancelled the Boeing contract after there had been delays of almost five years, and cost overruns approaching $5 billion (billion, not million).[17] Whatever problems Ball Aerospace has had with cost performance, they pale in comparison with Boeing's problems on FIA. So, in light of this monumental failure to manage cost on the part of Boeing, why did Ball Aerospace's cost performance issues, minor by comparison with Boeing, get cited as the reason for awarding the Ares I IUA contract to Boeing? Only Doug Cooke knows for sure.

Looking back at other things that were said both publicly and privately about what NASA wanted to do with this procurement, what happened to the thinking that they

- didn't want the same contractor for both the Upper Stage and the IUA,
- wanted new blood,
- wanted creativity and innovation, and

---

[17] Global Security, "Future Imagery Architecture (FIA)—2005 Restructuring," accessed April 25, 2019, https://www.globalsecurity.org/intell/systems/fia-2005.htm.

— wanted to be able to migrate aircraft avionics approaches and innovations into the IUA world?

Since Doug said that he had two essentially equal proposals to pick from, given his broad discretionary power as an SSO, he could have picked either one. Had he picked Ball Aerospace, he would have satisfied all four of the above-stated NASA objectives. And he could have cited Boeing's performance on FIA as the reason that he went with Ball Aerospace. But, for whatever reason, he didn't.

The truth of the matter is that Doug Cooke (legally) did no wrong. His selection appeared, and still appears, to have been within the scope of his very broad authority, and his rationale was certainly logical, at least if he didn't know about the problems that Boeing had had managing cost on programs like FIA. But it would have been just as easy to have picked Ball Aerospace and thus gotten new blood, a different contractor for the IUA than for the Upper Stage, and brand-new aircraft industry innovations that could have been brought into the human space flight world. As I said earlier, only Doug Cooke really knows why he did what he did, and he's not talking.

The Debriefing

As I mentioned earlier, we immediately requested a debriefing after the selection was announced, and it had gotten scheduled for Wednesday, December 19. We talked to Ball Aerospace's legal department, and they confirmed that we would have five days after the debriefing to file a protest, if we wanted to do that. Ball Aerospace, totally unlike Boeing, has historically been protest averse, and such filings by Ball Aerospace are rare. However, the magnitude of this pursuit was unprecedented in Ball Aerospace's history, so I wanted to preserve the possibility to protest by making as much time available to make a decision as I could. With the debriefing scheduled for December 19, this meant that we would have to do whatever we were going to do by the close of business five days later, which was December 24. Having worked in the government, I knew there was a good possibility that the government would be granted Christmas Eve off by presidential executive order, as that frequently happened when Christmas Eve occurred on a Monday or a Friday. Thus, practically speaking, we would have to make our decision by close of business on Friday, December 21. So, when I was trying to get the source selection statement out of Steve Beale at Marshall, I also requested that the debriefing be moved earlier, possibly to December 17. He agreed to discuss this with his team, but it only got moved to December 18. At least that was one additional day.

A debriefing is something winners and losers request after a selection decision—losers to see what they did wrong and if they might have grounds for a protest, and winners to see what they could have done better. It is typically a good process, and one most contractors avail themselves of. When a contractor gets debriefed by the SEB, which would be the case for the Ares I IUA, the debriefing is focused on what that contractor's evaluation was, i.e., you don't get to learn anything significant about what your competition did to win, beyond seeing how many significant strengths/strengths they had. So our debriefing would be principally about the SEB's evaluation of Ball Aerospace's proposal.

Eight of us, including one person each from our teammates, Hamilton Sundstrand and Pratt & Whitney Rocketdyne, arrived at Marshall in time for the start of our debriefing at one o'clock. The SEB was present, as was Steve Beale's deputy, Bryon Butler, who was slated to take over for Steve after his impending retirement. The environment in the room was, quite frankly, uptight. I don't know whether this was just nervousness, or because they were afraid that they would say something wrong, that might lead us to protest. I should note that the rules of engagement allowed us to ask questions and we did. Of course, anything of substance about our competition was off-limits.

Summarizing the results of the debriefing:

- They had a hard time explaining how they determined that we had significant past performance weaknesses on a few individual programs yet still gave us a "very good" rating overall; the issues with a few individual programs seemed to have been based more on talking to people than on analyzing actual cost performance data; this was a worrisome discussion.
- I brought up the subject of Boeing's past performance, asking whether they had used the exact same approach when they evaluated Boeing's past performance; I felt my question was general enough that I should have been able to ask it, but the

SEB chairperson, Johnny Stephenson, was not happy that I had asked this at all; nonetheless, eventually he said yes.

— I pressed my luck and went a bit further and asked if they had specifically used the FIA program in their evaluation of Boeing's past performance. Johnny got a bit agitated and said this question was off-limits, and that was that.

— We learned that our SAF approach had been the most significant strength of any proposer.

— Making me feel a lot better, we also learned that had we not added money to fix the significant weakness in the SAF cost-reduction area, that they would have added more money than we did to our most probable cost, and as I had surmised, we would have still had the weakness. I felt totally exonerated with respect to my decision about this issue, since at least one of my vice president colleagues had thought that had been a bad decision.

— We got much more detailed information about our significant strengths, and there were many things that we discovered we had done extremely well, which was quite rewarding; one of the most humorous discoveries in this area was that our day-one staffing had been a significant strength; I'm sure my colleague vice presidents loved that one.

I didn't really see anything that I felt warranted Ball Aerospace filing a protest, other than to just throw sand in the gears of the program, which I did not think would be a good thing for Ball Aerospace to do in the long haul.

We put all this together in Huntsville and then, the next day, December 19, had a special session by telecom with Dave Taylor and his staff to go over everything we had learned.

In this discussion, I recommended to Dave that we not protest, as I saw no reasonable grounds to do so. The bottom line was that Doug Cooke had two great proposals to pick from, and he could only pick one. I further recommended that Dave and I do coordinated phone calls to Mike Griffin (from Dave) and Dave King (from me) to tell them that we had looked at the source selection statement and received the debrief, and we felt it had been a fair competition. We would not protest so their program could continue to move forward, and we would continue to help NASA be successful in other ways. These calls took place an hour apart the next morning, Thursday, December 20. Obviously, these phone calls were well received.

So now it was really over. And time to go home for the holidays.

A Final Meeting with Robert Lightfoot

Before clearing out of Huntsville to go home for the holidays, I asked Parker Counts to see if he could get me in to see Robert Lightfoot immediately after the holidays. I had already talked to Dave King by telephone, so I saw no reason to revisit the subject of the Ares I IUA with him. Parker got this set up for Monday morning, January 7, 2008, in Robert's office at Marshall. I said a lot of the same things to Robert that I had said to Dave King about this having been a fair competition, that we wanted to continue to work closely with NASA. I also confirmed to Robert that Ball Aerospace intended to stay in Huntsville as we saw it as a good place to do business, with lots of opportunities, including many at Marshall. I also indicated that we were already working with Boeing to see if we could help them with this job given how much work that they now had on-going with Marshall, since they had won both the Upper Stage and the IUA. I did manage to get a smile out of Robert on that one.

Then I asked him if he had any lessons learned for us from this pursuit that he could share that would possibly help us in the future. What followed served to clearly substantiate the fact that Marshall senior management wasn't much involved in the decision process, and apparently, had only made the people available to prepare the RFP and provided the SEB to do the formal evaluation. Neither he nor Dave King were involved in the decision process (this blew me away, as that had not been how it worked when I was the Goddard deputy center director just a little over three years earlier).

When the final SEB package left Marshall to go to Washington, Steve Beale had said to Robert that it was the first time in his extensive experience at Marshall that he didn't feel he knew what the selection outcome was going to be (because it was such a close call). On December 11, at 8:30 on the night before the day of the selection announcement, he had gotten a call from Steve Beale saying the meeting with Doug Cooke had just ended, and that the answer was Boeing.

After that, he told me he thought we had done a great job; that he had been surprised when we made it through the down select gate, but that he then realized we were a serious contender.

We had some idle chitchat and then went our separate ways. I saw Robert briefly at the National Space Symposium in Colorado Springs in March 2009 (some one and a half years later), but haven't seen or talked to Dave King (who has now left NASA Marshall) since my call to him after the selection decision. In the meantime, and upon considerable reflection, it became increasingly clear that Marshall was still not in the inner circle of the management of the human space flight program as they had so very much desired, but that, instead, the Johnson Space Center remained very much in charge.

Post-decision Reflections

The Ares I IUA pursuit was a great experience for everyone involved, especially for Ball Aerospace as a company. We all grew a lot throughout this pursuit, especially given how far behind we were in the pack of competitors, at the time we got started. And many times we had to reach into our reserves, so to speak, to be able to do some pretty amazing things that we hadn't ever done before, but it seemed that we always came through when we had to. We went up against four of the really big boys, and beat three of them cleanly. And, despite the fact that our final competitor was Boeing, we produced a proposal that was good enough to have won, which was a wonderful thing to have been able to do.

But you can always learn something useful from almost any experience, and when a company spends as much money as we did on a losing pursuit, those lessons learned become so expensive, that you really need to heed them. So, with the benefit of perfect hindsight, what did we do well, and what could we have done better:

The good:

— Having our CEO Dave Taylor's undying commitment, support, and dedication to this pursuit was absolutely indispensable. He was our best cheerleader and supporter, and I can't count the number of times that he said, "Just get it done," to whoever was, at that moment, being an impediment to the quick resolution of an important issue.

— Assigning a senior vice president level person to lead this pursuit proved to have been essential; it didn't have to have been me, but the concept is a good one; it really serves to demonstrate a company's commitment to the pursuit in question. Plus, a leader at the vice president level can break down an awful lot of barriers to progress, that otherwise would simply fester as issues; not to mention isolating the team from everybody else, so that the team can focus on doing its jobs better.

— Declaring the Ares I IUA pursuit a must-win was absolutely vital; this allowed the pursuit to get corporate-level attention, and most importantly, corporate-level resources (key people, systems, priority, as well as funding).

— Establishing an office in Huntsville was also essential, at least with this customer. I don't believe we would have had any chance at all to win had we not done this.

— Moving the pursuit team to Huntsville was definitely the right move as well, as it would have been much, much harder to have written this proposal in Boulder, so much so that I do not believe that we would have had anywhere as good a proposal as we obviously did.

— Having a dedicated community liaison was super and served to do just what we needed to do (i.e., get us rapidly integrated into the business community).

— And I cannot say enough positive things about our pursuit team. It was a super bunch of folks who really came together and coalesced into a top-notch, high-performing group right before my eyes.

The less good:

— We needed to have done a better job of nurturing and growing our relationship with NASA Headquarters' senior leadership.

While we did meetings with NASA Headquarters senior leaders, participated in events in the DC area where NASA Headquarters leadership was present, and sponsored events at NASA Kennedy during space shuttle launches where NASA Headquarters leadership was present, the results of these interactions weren't what they needed to have been. Given that we weren't an incumbent in the human space flight area, we recognized early on that this would be a tough nut to crack. Thus, Carol Lane and her staff, Carolyn and I, plus some special consultants, met monthly to identify what else there was that we could be doing with NASA Headquarters. And we did uncover some new things to do, but the main impediment throughout, was that we simply didn't have convenient access to NASA Headquarters' senior leadership, due primarily to not being an incumbent in the human space flight area.

— We also could have done better in the key personnel area; we sometimes moved too slowly, and we weren't always able to bring the right people into the fold. A clear indicator of this is the difficulty that we had getting a senior executive and a program manager onboard. Each of these positions required people from outside Ball Aerospace with unique skills and experience that specifically addressed the requirements of this opportunity. Such people are not easy to locate, and are even harder to get onboard because of their high compensation expectations. The fact that we weren't able to get our first choice in either of these two cases was disappointing, to say the least. In the case of the senior executive level position, I ended up taking on the job, which was very disruptive to the company as evidenced by the reorganization that was required to make this happen. In the case of the program manager, the person that we ended up hiring for the job did bring unique and necessary skills and experience to the position, and had a solid track record as a program manager. But it took a lot of time to get these two key positions filled, and we should have accomplished that much more quickly, so that these two key folks could have begun influencing things much earlier.

- While we had been working on this for many months, it would also have helped had we been able to choose our teammates sooner, and thus formed our team sooner. I'm not actually sure that we could have accomplished this, and I was certainly very happy with the team that we ultimately established, but it did come pretty late in the process (i.e., at about the same time as the draft RFP release in April 2007).

- We needed better strategic intelligence about our competition. We did a black hat assessment of our competition and an independent price-to-win assessment, but these events were very negatively impacted by the scarcity of solid information about what our competition might do based on past history. Since the human space flight area was a new marketplace for us, we simply weren't as plugged in to our competition as we normally would have been.

There are certainly more good things, as well as some additional less good things that we did or didn't do, but I believe the above are the most important ones. Let me say again that we did extremely well, but like all things in life, you can always do better, especially the next time out.

On the NASA side of this procurement, my main disappointment was that the things that we, and the other bidders, were told by NASA that they wanted, such as new blood, creative/innovative approaches, two independent contractors for the Upper Stage and the IUA, ended up not being what they wanted at all. *Rather, it seemed NASA wanted the incumbent.* So, at the end of the day, it seemed to be business as usual, which meant a lot of time, effort, and money may have been wasted on this procurement by the losing bidders.

Clearly, if there's a single significant lesson learned from this procurement, it is to *"beware the incumbent!"*

Moving On

As Carolyn and I reflected on the loss, we decided we needed a break. I had been active in the aerospace industry for forty-five years, and the intensity of this pursuit had taken its toll. Carolyn had spent thirty-five years in the aerospace industry and felt the same way, so we decided to take a year off to smell the roses before coming back to try again. So we gave a month's notice, packed up our offices in Huntsville, packed up our offices in Boulder, and said goodbye to Ball Aerospace. It had been a great ride, but it was time to do something different.

I did an interview with *Aeronews*, the Ball Aerospace newsletter, before I left, where I was quoted as having said, "My forty-five years in aerospace had been a hoot." And it was. I particularly enjoyed being at Ball Aerospace for almost four of those forty-five years. In the *Aeronews* article I was also quoted as saying, "The men and women of Ball Aerospace are the freest thinking and most creative people I have ever had the privilege to work with." And they were, and that was especially the case for the Ares I IUA pursuit team.

Just before leaving Ball Aerospace, I got time with Dave Hoover, the CEO of Ball Corporation, to say goodbye and tell him I was very sorry that I hadn't been able to win the Ares I IUA pursuit for Ball. We had a very nice informal chat, with him even commenting on the fact that we were sitting in the exact same seats in his office when I had first interviewed with him, now some three and a half years earlier, but that it really didn't seem all that long ago. As we went on to discuss the loss, he consoled me by saying, "Hey, it's not your fault. It should be obvious that NASA simply picked the wrong team!" Like I said earlier, this is a person that you always want on your side.

So on Friday, January 11, 2008, Ball Aerospace threw a nice going-away party for Carolyn and me. Dave Taylor and most of his senior staff were there, and, of course, the usual storytelling and roasting took place in spades. Not the least of which was the retelling of the "clown car" story that we had gotten so much mileage out of while we were in Huntsville. It was all good fun.

And then we were gone.

Epilogue

It is now eleven years since we left Ball Aerospace, seven years since I did a significant revision of the original draft of this book, followed by minor editorial-type revisions in March 2017, and again in March 2019. At the time I completed the first draft of this book (February 2009), it had only been a year since we had left Ball Aerospace, so everything about the Ares I IUA pursuit was still very fresh in my mind. I wrote this book in the two-bedroom loft that we lived in while working for Ball Aerospace, which is located in Boulder, Colorado, near the base of the beautiful Rocky Mountains. And being there must have been good for me, as I really had no problem putting pen to paper, with the words flowing almost faster than I could type, especially since I am a "hunt and peck" typist anyway (imagine doing that for more than forty thousand words!).

After Carolyn and I left Ball Aerospace, a number of interesting things happened. We all had known that Boeing's plate would be beyond overloaded in Huntsville if they won both the Ares I Upper Stage and IUA contracts. So, as had been our plan in the event of a loss, Ball Aerospace approached Boeing immediately after the announcement that Boeing had won the IUA contract, to see if Ball Aerospace could help them with their IUA work. And the answer was yes, with Boeing initially contracting with Ball Aerospace to build and deliver twelve engineering development unit (EDU) flight computers to Boeing. And there was much more work being planned by Boeing for Ball Aerospace to do.

But a funny thing happened on the way to the bank, so to speak.

Barack Obama (a Democrat) got elected as the forty-fourth president of the United States, taking office in January 2009. Once he got his feet firmly on the ground, one of the many changes he made was to cancel President Bush's (a Republican) Vision for Space Exploration program, and President Obama then sought to put his own human space flight program in place, a program centered around the development of a new launch vehicle referred to as the Space Launch System (SLS) and intended to eventually be used to send humans to Mars. Additionally, the space shuttle program launched its 135th and last mission (on *Atlantis*) on July 8, 2011. So the United States found itself without any means to carry humans to low earth orbit (LEO) for the first time since April 12, 1981, when the space shuttle *Columbia* flew into Earth orbit for the first time. Thus, the United States became dependent on the Russians to ferry astronauts back and forth to the International Space Station (ISS), while US-based commercial companies such as SpaceX, and, yes, Boeing, worked on developing an independent US capability to carry humans to the ISS. While President Obama's SLS (and President Bush's Orion capsule) have made substantial progress, the first flight of SLS/Orion is not currently planned until 2020, and it is not planned to carry astronauts until 2022. Donald Trump (a Republican) was elected the forty-fifth president, taking office in January 2017. And guess what? His Administration's plans for space have been reoriented back to the moon once again. Thus, NASA's human space flight destination over the last fifteen years has not only been inconsistent, but has become highly politicized: moon ... Mars ... moon—is that any way to run a railroad to the stars?

However, in March 2019, Vice President Mike Pence charged NASA with accelerating its schedule to return to the moon from 2028 to 2024. One potential benefit of this accelerated schedule, if met, would be that it would become much more difficult for a future Administration to change the destination yet again. So maybe this will finally be the time that US astronauts again leave low earth orbit, after a hiatus of over fifty years. I certainly hope so.

After leaving Ball Aerospace, we took a year off, and that was great fun. We stopped getting up at five thirty, something that it seemed that we had been doing our entire lives. We skied in Colorado all that winter, where the great skiing was totally uninterrupted by work. Then we sailed the summer away on the Chesapeake Bay. We next took that long-delayed fortieth-anniversary trip (we had been working in Huntsville in September 2007, when the actual event occurred) for almost three weeks to the South of France, staying a week at a five hundred-year old farmhouse in Provence, slipping down to the Mediterranean for a few days, and going into and out of Paris, staying a few days in that lovely city on each end. Our daughter, Tiffany, spent her summer college break with us in Annapolis, where she taught sailing. And we got to see a lot of our son, Jason, who worked in nearby Washington, DC. We also saw a lot of college football, including visiting our daughter in Oregon for Dad's weekend at Oregon State after returning from France, and then we were into the holidays, which we enjoyed immensely. But our "loss" had been on my mind all throughout this wonderful year, so I knew that when winter came, it would be time to try to put this story down on paper, so that others could read it and hopefully learn from it. And of course, by so doing, perhaps I could finally have some closure and put the loss behind me.

With a first draft of the book written that winter, we turned our attention to what we were going to do next. While we had enjoyed our year off immensely, "the phone was still ringing," though we knew it wouldn't go on much longer. You can only tell people no so many times, and then they lose interest in asking you to do special projects for them.

So Carolyn and I formed Townsend Aerospace Consulting, LLC (TAC),[18] and we are now in our eleventh year of operating a woman-owned small business (yes, Carolyn is the principal owner, and therefore my boss!). TAC provides expert advice to the aerospace industry, and currently has two private-sector clients, through which TAC is doing work for both NASA and NOAA. Looking back, TAC's work over the last decade has ranged from performing an independent assessment of one of the NASA Goddard Space Flight Center's key human capital processes, to me being the chair of NOAA's standing review board for the development of their new GOES-R weather satellite system, to doing "color" team proposal reviews for some of our private-sector clients. It has been not only profitable for us to date but, perhaps more importantly, great fun. Additionally, I have also found time to give back to my community through pro bono work in support of various activities for the National

---

[18] To learn more about TAC, please go to www.townsendaerospace.com.

Academy of Sciences, most recently being a reviewer for the latest Earth Science Decadal Survey, a guideline document used by NASA, NOAA, and USGS to implement their Earth Science programs of the future. But working for TAC caused my book to be put on the back burner for quite a while; it only recently rose high enough on the priority list to get some much-needed focused attention such that my book is finally done!

At this point, some twelve years down the road from the events in Huntsville, all I hope is that, for those of you who are not in the aerospace industry who read this book, you will have your eyes opened to what really goes on behind the scenes when a key aerospace industry contract award is announced. For those of you who are in the aerospace industry, perhaps there is a thing or two that you can learn from my telling of this story. And for those of you from NASA, I sincerely hope that some of you will be motivated to try to develop some lessons learned for NASA, such that the substantial waste of money that occurred with this procurement might be avoided, or at least minimized, in the future. In that regard, I would say to NASA, if you don't really want new blood, innovation, and creativity, then don't raise expectations unnecessarily by asking for it. I say this because my true love (other than Carolyn) has, and will always be, NASA, and I just want to see NASA do better than what I saw NASA do with the Ares I IUA procurement.

So, with that, I wish each of you well.

Bill Townsend
Boulder, Colorado
July 2019

Appendix 1

Huntsville, Alabama: The Job Location

Huntsville, Alabama[19,20]

Huntsville is located in North Alabama in the heart of the Tennessee River valley. The origin of Huntsville, Alabama, is credited to John Hunt, a veteran of the Revolutionary War. He came to Alabama from Tennessee in 1805 when Alabama was still a portion of the Mississippi territory that the Indians called Ah-la-bama. John Hunt built his log cabin in the beautiful wooded valley just above the "big spring," thus founding the town that would bear his name.

In 1819, Alabama was made a state, and Huntsville was chosen as its first Capital.

The Civil War and Reconstruction period brought hard times to Huntsville. By the turn of the century, emphasis was placed on industry, and the scars of the Civil War began to heal.

Huntsville expanded with the onset of World War I, but industry declined during the Depression in the 1930s. During these years, Huntsville became famed as the "watercress capital of the world," and

---

[19] Wikipedia contributors, "Huntsville, Alabama," accessed February 1, 2009, https://en.wikipedia.org/wiki/Huntsville,_Alabama.

[20] Huntsville, "History of Huntsville," accessed April 28, 2019, https://www.huntsvilleal.gov/business/city-of-huntsville/the-history-of-huntsville/.

Madison County, which surrounds Huntsville, was Alabama's leader in cotton production.

Just before the United States entered World War II, Huntsville Arsenal was constructed adjacent to the city to manufacture chemical artillery shells. A few months later, another plant was constructed to assemble explosives for the shells. During the peak of World War II production, the two arsenals—Huntsville and Redstone—employed about twenty thousand people.

During the immediate postwar period, arsenal activities were sharply curtailed. In 1950, the army transferred its small group of missile experts, including Dr. Wernher von Braun, to Huntsville from Fort Bliss, Texas.

In 1958, this army team put aloft a Jupiter C missile. As a result, the complete US Army missile development and training program, was transferred to the newly established NASA Marshall Space Flight Center in Huntsville.

In 2007, Huntsville was the largest city in northern Alabama with the city proper having about 170,000 residents, and the Huntsville metropolitan area having almost 400,000 residents. Huntsville is home to both the Army's Redstone Arsenal and NASA's Marshall Space Flight Center, and is nicknamed Rocket City for its close history with US space missions.

Huntsville's economy was nearly crippled, and growth came to a near standstill in the 1970s, following the closure of the Apollo program, but the emergence of the space shuttle and the ever-expanding field of missile defense in the 1980s helped give Huntsville a resurgence that continues to this day. The city continues to be the center of rocket-propulsion research in the United States and is home to large branches of many defense contractors.

Huntsville has a humid subtropical climate. It experiences hot, humid summers and generally mild winters, with average high temperatures ranging from 89°F in the summer to 49°F during winter. Some years, Huntsville experiences tornadoes during the spring and fall. While most winters have some measurable snow, significant snow is rare in Huntsville.

Huntsville has three official historic districts—Twickenham, Old Town, and Five Points. A historic house tour is held every spring.

Other Huntsville highlights of note (most which were referred to earlier in this book) include

- The US Space & Rocket Center, which is home to the US Space Camp. Additionally, it houses the only Saturn V rocket designated as a national historic landmark.
- The Von Braun Center, located adjacent to Big Spring International Park, opened in 1975 and has an arena capable of seating ten thousand, a two thousand-seat concert hall, a five hundred-seat playhouse, and 150,000 square feet of convention space.
- The Huntsville Museum of Art, which is in Big Spring International Park, offers permanent displays, traveling exhibitions, and educational programs for children and adults.
- The Huntsville Symphony Orchestra, founded in 1955 in response to the arrival of Wernher von Braun, is Alabama's oldest, continuously operating professional symphony orchestra, featuring high-quality performances of classical, pops, and family concerts, and extensive music education programs serving public schools.
- The Huntsville Botanical Garden features educational programs, woodland paths, broad grassy meadows, and stunning floral collections.
- The University of Alabama in Huntsville is the largest university serving the greater Huntsville area. The research-intensive university had more than 7,200 students in 2007. Approximately half of the university's graduates earn a degree in engineering or science, making the university one of the largest producers of engineers and physical scientists in Alabama.
- The Huntsville International Airport is served by several regional and national carriers and offers nonstop flights to many airports across the eastern United States. However, Huntsville International gets its name because of its reputation as a cargo transport hub. Many delivery companies have hubs in Huntsville, making delivery flights to Europe, Asia, and Mexico.

- The *Huntsville Times* is Huntsville's only daily newspaper and in 2007 had a weekday circulation of sixty thousand, which rose to eighty thousand on Sundays.

Economically, the area was almost totally driven by the defense and space industries, which awarded contracts that totaled in excess of $25 billion in 2007.

So, as you can see, Huntsville is a moderate-size community that is sophisticated, vibrant, and has lots to offer those who live there or are thinking of moving there. Huntsville also prides itself on being a leader in a number of areas, and in the 2007 timeframe, sported top ten rankings by various magazines and other publications in many areas, for example:

- America's number-four place to live and work
- One of the top-ten metro areas for business vitality
- One of the top-five cities for professional workers
- One of the top-ten metro areas for scientists and engineers
- One of the country's fifteen great cities for job seekers
- One of the top-ten cities for job growth and affordability
- America's number-one small city of the future

NASA Marshall Space Flight Center[21,22]

Located in Huntsville, Alabama, Marshall was established in 1960, and Dr. Wernher von Braun became its first center director. Von Braun was one of the most important rocket developers and champions of space exploration during the period between the 1930s and the '70s. As a youth, he became enamored with the possibilities of space exploration by reading the science fiction of Jules Verne and H. G. Wells. As a means of furthering his desire to build large and capable rockets, in 1932 he went to work for the German army to develop ballistic missiles.

Von Braun is well known as the leader of what has been called the

[21] Wikipedia contributors, "Marshall Space Flight Center," accessed February 1, 2009, https://en.wikipedia.org/w/index.php?title=Marshall_Space_Flight_Center&oldid=891984341.

[22] NASA MSFC, "Historical Facts," accessed April 28, 2019, https://history.msfc.nasa.gov/history_fact_sheet.html.

"rocket team," which developed the V-2 ballistic missile for the Nazis during World War II. The brainchild of von Braun's rocket team operating at a secret laboratory at Peenemünde on the Baltic Coast, the V-2 rocket was the immediate antecedent of those used in space exploration programs in the United States and the Soviet Union. Before the Allied capture of the V-2 rocket complex, von Braun engineered the surrender of five hundred of his top rocket scientists, along with plans and test vehicles, to the Americans. As part of a military operation called Project Paperclip, he and his rocket team were scooped up from defeated Germany and sent to America, where they were installed at Fort Bliss, Texas. There they worked on rockets for the US Army, launching them at White Sands Proving Ground, New Mexico. In 1950, von Braun's team moved to the army's Redstone Arsenal near Huntsville, Alabama, where they built the army's PGM-19 Jupiter ballistic missile.

In 1960, his rocket development center transferred from the army to the newly established NASA Marshall Space Flight Center. In 1961, when President John F. Kennedy envisioned an American on the moon by the end of the decade, NASA turned to Marshall Space Flight Center to create the incredibly powerful rocket, the Saturn V, needed to turn this presidential vision into reality. Von Braun was the principal architect of the Saturn V.

Today, NASA is working on America's latest space vision: development of the largest launch vehicle ever built, the space launch system (SLS), and together with the Orion capsule, sending astronauts to the moon ... or is it Mars? In either case, once again, NASA has turned to Marshall for the development of the powerful rocket known as SLS, which is planned to succeed the space shuttle.

Regarding President Bush's original Vision for Space Exploration, Marshall had the responsibility to procure the Ares I Instrument Unit Avionics (IUA). As noted earlier, the general IUA approach has been drawn directly from the approach that was used by von Braun to develop the Saturn V Instrument Unit (IU).

Appendix 2

Ball Aerospace: The Company[23,24]

Ball Aerospace & Technologies Corp., or Ball Aerospace for short, is a medium-size aerospace industry company located in Boulder, Colorado. It is a wholly owned subsidiary of Ball Corporation, a major global packaging company located in Broomfield, Colorado, with sales in excess of $9 billion annually in 2007. If the name Ball sounds familiar, it's because they used to make Ball canning jars, something your mother or grandmother may have used when you were growing up. Ball Corporation is a publicly traded company (BLL) and manufactures high-quality metal and plastic packaging, primarily for beverages, foods, and household products, as well as aerospace and other technologies and services, primarily for the US government. In 2007, Ball Corporation celebrated its 135th anniversary.

Ball Aerospace is one of five operating segments of Ball Corporation and the only one that has nothing to do with packaging. In 2007, its sales were around $800 million, or almost 10 percent of its parent corporation. Ball Aerospace supports critical missions of important national agencies such as the Department of Defense, NASA, NOAA, and other US

---

[23] Ball Corporation, "About Ball," accessed January 29, 2009, https://www.ball.com/na/about-ball.

[24] Wikipedia contributors, "Ball Aerospace & Technologies," accessed April 29, 2019, https://en.wikipedia.org/w/index.php?title=Ball_Aerospace_%26_Technologies&oldid=866627811.

government and commercial entities. The company develops and manufactures spacecraft, advanced instruments and sensors, components, data exploitation systems, and RF solutions for strategic, tactical, and scientific applications. In 2007, Ball Aerospace had approximately three thousand employees and had recently celebrated its fiftieth anniversary. It is the oldest-operating segment of Ball Corporation and is known throughout the aerospace industry as a technology innovator.

Established in 1956, two years before NASA was established, Ball Aerospace has produced numerous scientific and technological firsts. Here are a few examples:

- First solar observation satellite: The Orbiting Solar Observatory (OSO), launched in 1962, was the first spacecraft built at Ball Aerospace and one of seven such spacecraft the company built for NASA. The OSO returned data on the ultraviolet, x-ray, and gamma ray emissions of the sun and our galaxy.
- First instrument to provide all-sky imaging of the universe in infrared from space: The Infrared Astronomical Satellite (IRAS), launched in 1983, was the first space-based, long-life, cryogenically cooled infrared telescope. Ball Aerospace built the IRAS instrument, which contained superfluid helium to cool an infrared sensor to nearly -450 degrees Fahrenheit.
- First instrument to provide confirmation of the Antarctic ozone hole: The Solar Backscatter Ultraviolet Radiometer (SBUV/2) helped confirm the ozone hole above Antarctica in 1987. The company built nine SBUV/2s between 1984 and 2002.
- First to create the highest submeter resolution on a satellite for commercial remote sensing: Ball Aerospace designed, built, integrated, and tested the spacecraft and the sixty-one-centimeter imaging system that comprises the QuickBird remote-sensing system. Launched in 2001, QuickBird was the highest-resolution commercial satellite in operation in Earth's orbit in 2007.
- First spacecraft to intercept a comet: Ball Aerospace designed, built, and tested the two spacecraft that comprised the Deep Impact mission to collide with Comet Tempel 1 on July 4, 2005,

when nearly eighty-three million miles from Earth, giving scientists a look into its composition and structure.

- First instrument to return high-resolution images of the Mars surface from an orbiting platform: Ball Aerospace's High-Resolution Imaging Science Instrument (HiRISE) was launched aboard the Mars Reconnaissance Orbiter (MRO) in 2005 to provide color stereo images of the Mars surface at six times higher resolution than any existing images.

- First spacecraft to provide mapping of the Earth's total ice volume: Under its Rapid Spacecraft Development program, NASA selected Ball Aerospace to build the ICESat spacecraft. ICESat, launched in 2003, uses the Geoscience Laser Altimeter System to measure changes in the thickness of ice sheets in Antarctica and Greenland.

- First spacecraft designed for on-orbit, autonomous servicing: Ball Aerospace built the Next Generation Satellite and Commodities Spacecraft (NEXTSat/CSC) and its ground support equipment and participated in the 2007 launch and mission operations. The spacecraft is part of the Orbital Express Advanced Technology Demonstration Program intended to prove techniques for on-orbit refueling, reconfiguration, and repair of spacecraft in orbit.

Aside from these most notable past firsts, Ball Aerospace continues today to be engaged in further cutting-edge space work, such as the following:

- First instrument to return high-resolution images of Pluto: Ball Aerospace developed the Ralph imager for the New Horizons mission to Pluto, launched in 2006. In 2015, Ralph provided images to create maps of Pluto, its moon, Charon, and other Kuiper Belt objects.

- First spacecraft to search for extrasolar terrestrial planets from deep space: Ball Aerospace built and tested the spacecraft and photometer for the Kepler mission. Launched in 2009, Kepler has steadily been discovering potentially habitable planets around distant stars.

- First optical system expected to study objects four hundred times fainter than any current telescope: Ball Aerospace is developing the advanced optical technology and lightweight mirror system at the heart of the James Webb Space Telescope, expected to be launched in 2021. The diameter of Webb's primary mirror will measure approximately 6.5 meters. The primary mirror is comprised of eighteen hexagonal mirror segments, each approximately 1.3 meters in size. The individual mirror segments are phased in space using computer-controlled actuators that can adjust the position and shape of the mirrors to give the telescope a high-quality sharp image. Because the Webb is an infrared telescope, the mirrors and actuators must function at temperatures as low as -400 degrees Fahrenheit.
- And as the icing on the cake, so to speak, Ball also built the Corrective Optics Space Telescope Axial Replacement (COSTAR), the supplemental optics installed into the Hubble Space Telescope in 1993 during the first Hubble servicing mission, that successfully corrected the spherical aberration of Hubble's primary mirror.

Continuing to support NASA in its Hubble endeavors, when NASA launched the space shuttle in May 2009 to service the Hubble Space Telescope for the fifth time, it included the installation of two brand-new instruments built by Ball Aerospace:

- The Cosmic Origins Spectrograph (COS): Built by Ball Aerospace to study fundamental problems in cosmology and astrophysics.
- The Wide Field Camera 3 (WFC3): Ball Aerospace developed the optical assemblies, instrument electronics, and detectors for WFC3 to improve Hubble's discovery efficiency by a factor of approximately thirty and to extend its outstanding imaging performance.

Given the installation of these two instruments into Hubble in 2009, all the current Hubble instruments will have been built by Ball Aerospace—quite an amazing accomplishment and a clear indicator of the respect that NASA has for the capabilities of Ball Aerospace.

Many of the above activities, including Deep Impact, Kepler, the James Webb Space Telescope, HiRISE on the Mars Reconnaissance Orbiter (MRO), Ralph on the New Horizons mission, and the two new Hubble Space Telescope instruments, were active in the Civil Space Systems Business Unit during my tenure as its vice president and general manager. I can't even begin to describe how proud I am of the excellent work done in these areas by the dedicated men and women of Ball Aerospace. But, having said that, I felt it was past time to turn our attention to a new NASA marketplace—the NASA Exploration program—which I thought was an equally excellent match for the creative and innovative talents of Ball Aerospace, as NASA's science area had been for virtually the entire fifty-plus-year history of Ball Aerospace. Thus, shortly after my arrival at Ball Aerospace, we moved into NASA's Exploration area and, more specifically, the Ares I IUA competition.

Printed in the United States
By Bookmasters